BASIC
ELECTRONIC
AND
ELECTRICAL
DRAFTING

JAMES D. BETHUNE

Wentworth Institute of Technology
Boston, Massachusetts

BASIC ELECTRONIC AND ELECTRICAL DRAFTING

PRENTICE-HALL, INC., *Englewood Cliffs, New Jersey 07632*

Library of Congress Cataloging in Publication Data

Bethune, James D.
   Basic electronic and electrical drafting.

   Includes index.
   1. Electronic drafting.   2. Electric drafting.
I. Title.
TK7866.B49 1980   621.3'022'1   79-26803
ISBN   0-13-060301-5

Editorial/production supervision and
interior design by *Virginia Huebner*
Page layout by *Gail Collis*
Cover design by *Edsal Enterprises*
Manufacturing buyer: *Gordon Osbourne*

Printed in the United States of America

10 9 8 7 6 5 4 3 2 1

PRENTICE-HALL INTERNATIONAL, INC., *London*
PRENTICE-HALL OF AUSTRALIA PTY. LIMITED, *Sydney*
PRENTICE-HALL OF CANADA, LTD., *Toronto*
PRENTICE-HALL OF INDIA PRIVATE LIMITED, *New Delhi*
PRENTICE-HALL OF JAPAN, INC., *Tokyo*
PRENTICE-HALL OF SOUTHEAST ASIA PTE. LTD., *Singapore*
WHITEHALL BOOKS LIMITED, *Wellington, New Zealand*

To KENDRA

CONTENTS

chapter three
SCHEMATIC DIAGRAMS
49

chapter four
CONNECTION DIAGRAMS
67

chapter five
BLOCK AND LOGIC DIAGRAMS
79

chapter six
PC BOARD DRAWINGS
91

chapter seven
CHASSIS DRAWINGS

103

chapter eight
PICTORIAL DRAWINGS

119

chapter nine
GRAPHS AND CHARTS

137

chapter ten
RESIDENTIAL ELECTRICAL WIRING

159

chapter eleven
INDUSTRIAL WIRING DIAGRAMS

173

APPENDIX

197

INDEX

205

PREFACE

Courses in electronic/electrical drafting usually face three problems; large class sizes, limited class hours, and an extensive amount of subject material. These problems are further complicated by the fact that much of the subject material deals with concepts not yet fully understood by students. For example, students must learn how to prepare schematic diagrams before they have completed courses in circuit theory and design.

This book has been written with these problems in mind. The book emphasizes the drawing aspects of electronic/electrical drafting (some basic theory and design applications have been included) and supplements the textual material with many illustrations and solved sample problems to enable the instructor to present the subject material in lecture form and then assign problems. The student, in turn, solves the problems by referring to lecture notes and by following solved example problems, procedures, and illustrations presented in the text. Thus, the student can optimize board time which, it is to be hoped, will maximize learning.

The first chapter of the book is an extensive review of basic drafting fundamentals. It has been included for courses that must teach both basic and electronic/electrical drafting. The material may also be helpful to students who need to review basic drafting before proceeding with electronic/electrical drafting.

Adopting any new text is at best inconvenient. The instructor must work out all the exercise problems, determine which are easy and which are difficult, and then experiment with the problems to see how long students take to complete them. To help minimize this inconvenience, a detailed solutions manual has been prepared for this book. It contains the solutions for almost all the exercise problems as well as a difficulty rating and an estimated completion time for each problem. Some sample quizzes and exams have also been included.

Several people deserve special recognition for their help in the development of this book: Cary Baker, Dave Boelio, and Virginia

Huebner, my editors at Prentice-Hall, who were always willing to answer my questions and supply good counsel; and Chris Duncombe, whose meticulous attention to detail has resulted in the clear, easy-to-understand photographs used in the book. All photographs not specifically credited to other sources, were taken by Chris. Elaine Beaupre typed the manuscript and corrected my miserable spelling. Thanks to each of you.

JAMES D. BETHUNE

Boston, Massachusetts

BASIC
DRAFTING
REVIEW

1-1 INTRODUCTION

This chapter is a review of the fundamentals of basic drafting. It represents only a brief outline of the subject matter. The student requiring a more extensive review is referred to one of the following drafting texts:

Bethune, J. D., *Essentials of Drafting*, 1977. Englewood Cliffs, N.J.: Prentice-Hall, Inc.

Luzadder, W. J., *Fundamentals of Engineering Drawing*, 7th ed., 1977. Englewood Cliffs, N.J.: Prentice-Hall, Inc.

McCabe, F. C., et al. *Mechanical Drafting Essentials*, 4th ed., 1967, Englewood Cliffs, N.J.: Prentice-Hall, Inc.

1-2 DRAFTING EQUIPMENT

Figure 1-1 pictures most of the drafting equipment needed to prepare electronic drawings and diagrams: a T-square, 30-60-90 triangle, 45-45-90 triangle, lead holder, eraser, erasing shield, protractor, scale, drawing tape, circle template, compass, and an electronic symbol template. In the upper right corner of the drawing board is a lead sharpener and a styrofoam block. The styrofoam block is used to remove excessive graphite from a freshly sharpened lead.

Figure 1-2 shows a drafting machine which replaces a T-square, triangles, scale, and protractor. Operating instructions for drafting machine are supplied by manufacturers.

Figure 1-3 shows how a T-square and triangles are used as guides when drawing horizontal and vertical lines. Remember to pull the lead-

3

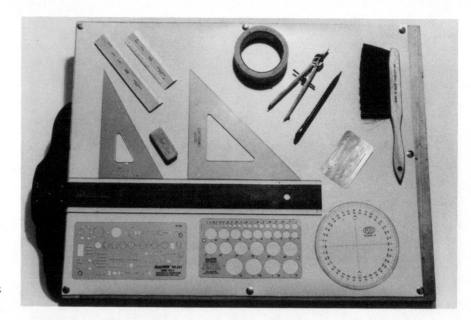

Figure 1-1 Basic drawing equipment needed to prepare electronic drawings.

Figure 1-2 Using a drafting machine. (*Photograph courtesy of Teledyne Post, Des Plaines, Illinois 60016.*)

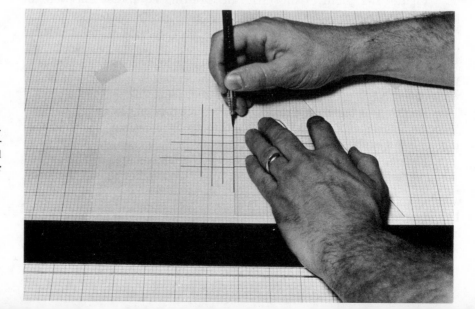

Figure 1-3 Using a T-square and triangle. The T-square is used as a guide for drawing horizontal lines and the T-square and triangle are used as shown as a guide for drawing vertical lines.

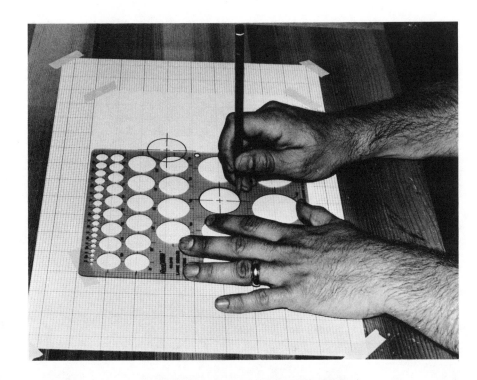

Figure 1-4 Using a circle template. Remember to keep the pencil vertical.

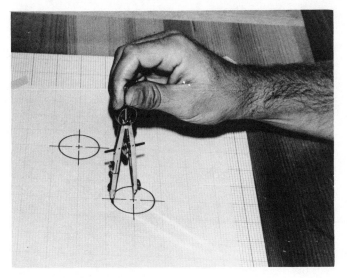

Figure 1-5 Using a compass.

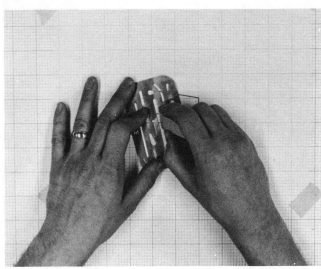

Figure 1-6 Using an erasing shield.

holder across the paper as pushing can result in a torn paper. Figure 1-4 shows how to use a circle template, Figure 1-5 shows how to use a compass, and Figure 1-6 shows how to use an erasing shield.

1-3 LINES

Many kinds of lines are commonly used on technical drawings: visible, hidden, center, leader, and phantom to name but a few. Figure 1-7 illustrates several types. Lines used for dimensioning are illustrated in Figure 1-8.

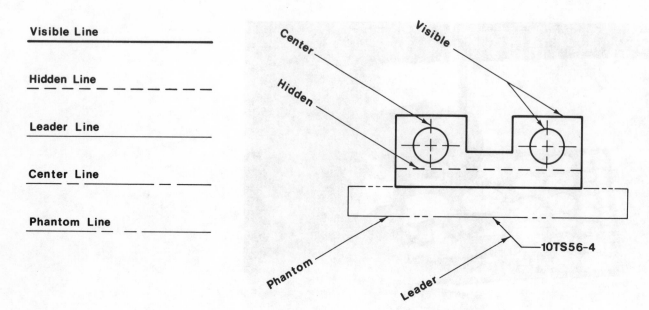

Figure 1-7 The different kinds of lines used on technical drawings.

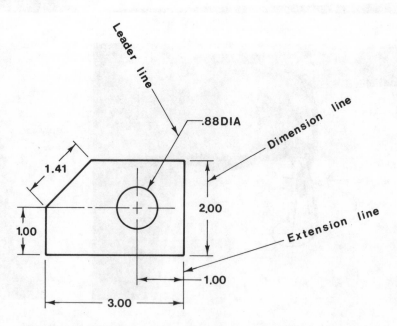

Figure 1-8 The different kinds of lines used for adding dimensions to a technical drawing.

1-4 LETTERING

Figures 1-9 and 1-10 show the shape and style of the letters and numbers most often used on technical drawings. Either the vertical or inclined style is acceptable although the two styles should never be mixed. The most widely accepted height for letters and numbers is ⅛ (.13) to 3⁄16 (.19).

ABCDEFGHIJKLMNOPQRSTUVWXYZ

abcdefghijklmnopqrstuvwxyz

0123456789

Figure 1-9 Vertical style letters.

ABCDEFGHIJKLMNOPQRSTUVWXYZ

abcdefghijklmnopqrstuvwxyz

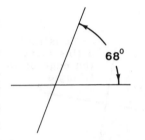

0123456789

Figure 1-10 Inclined style letters.

To help insure even lettering, draftsmen use guidelines as shown in Figure 1-11. It is acceptable to leave guidelines on the drawing—they need not be erased after the lettering has been completed.

Figure 1-11 Use guide lines (very light horizontal lines) to help insure even lettering.

1-5 GEOMETRIC CONSTRUCTIONS

This section includes six geometric constructions which are often required when creating electronic drawings. They are presented in step-by-step format in Figures 1-12, 1-13, 1-14, 1-15, 1-16, and 1-17.

1-6 ORTHOGRAPHIC VIEWS

Orthographic views are two-dimensional drawings which are used to define three-dimensional shapes. They are used in lieu of three-dimensional pictures because they permit much greater drawing ac-

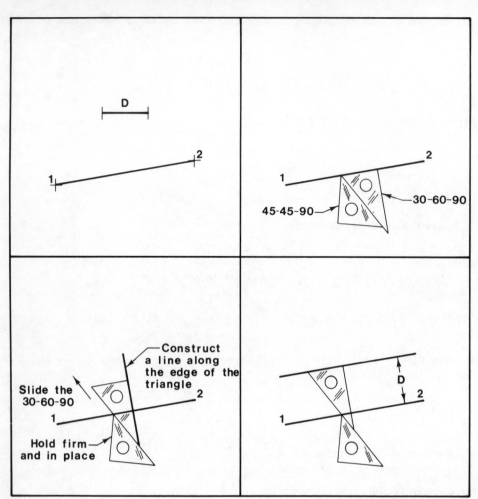

Figure 1-12 How to draw parallel lines using two triangles.

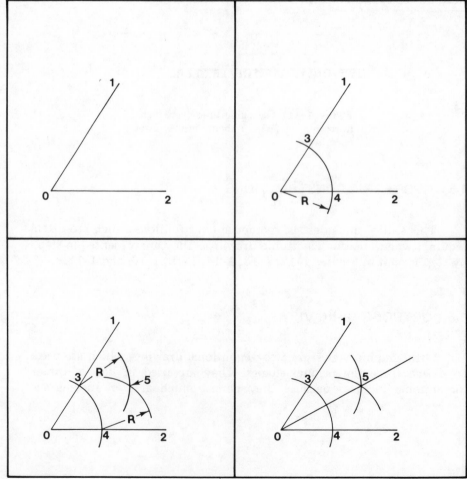

Figure 1-13 How to bisect an angle.

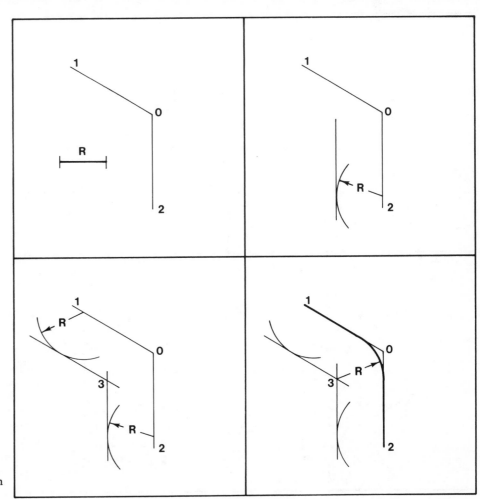

Figure 1-15 How to draw a fillet to an obtuse angle.

Figure 1-14 How to divide a line into any number of equal spaces.

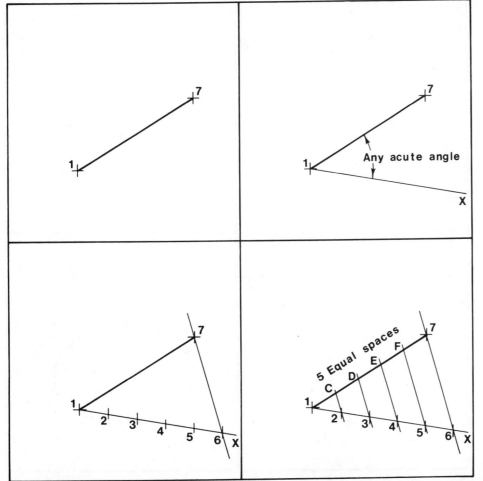

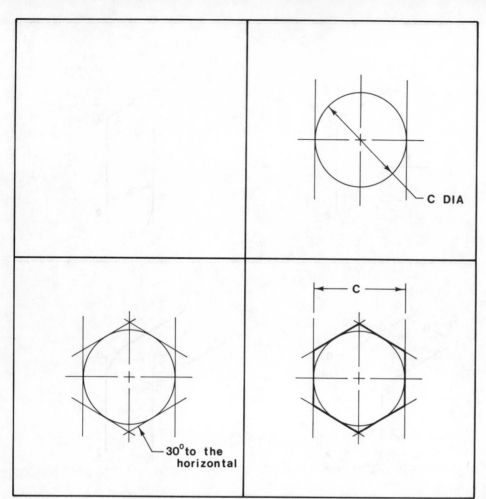

Figure 1-16 How to draw a hexagon for a given distance across the flats.

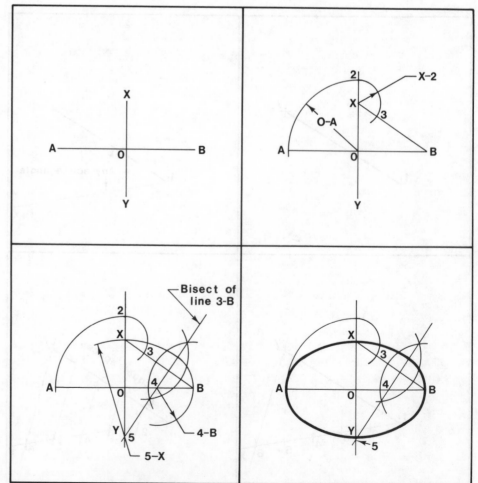

Figure 1-17 How to draw an ellipse for a given major and minor axis.

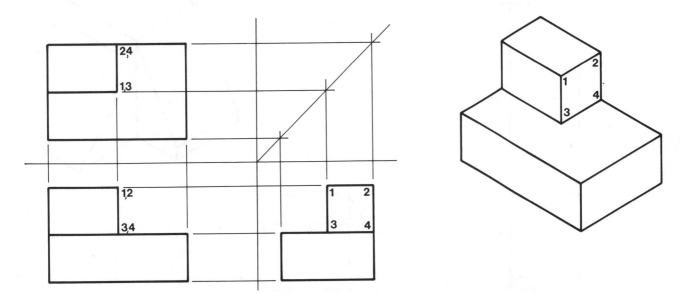

Figure 1-18 An object and its front, top, and right side ortho-graphic views.

Figure 1-19 An object and its orthographic views which include hidden lines.

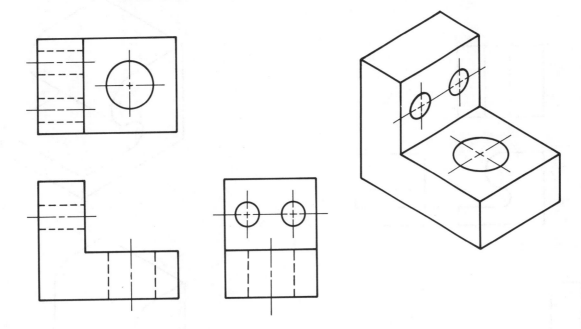

curacy. Figure 1-18 illustrates an object together with the front, top, and right side orthographic views of the object.

Hidden lines are used to picture surfaces and edges which are not directly visible in orthographic views. Figure 1-19 pictures an object whose orthographic views contain hidden lines.

Inclined surfaces are drawn orthographically as shown in Figure 1-20 and rounded surfaces are drawn as shown in Figure 1-21. In each

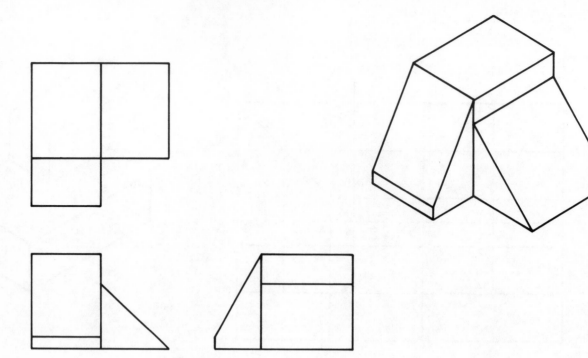

Figure 1-20 How to draw orthographic views of inclined surfaces.

Figure 1-21 How to draw orthographic views of rounded surfaces.

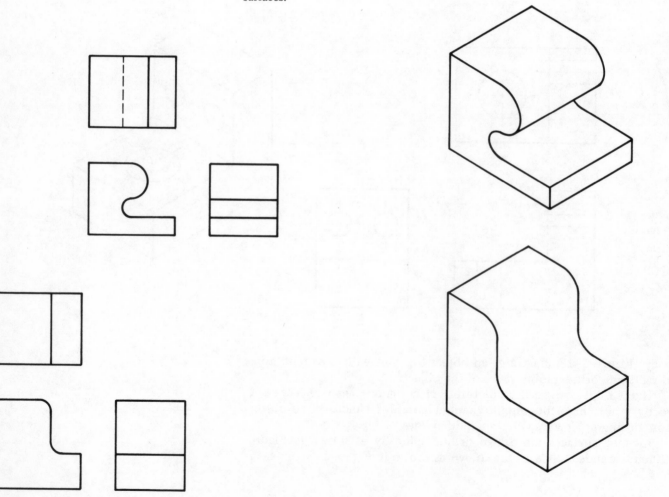

12

case only the profile orthographic views show the true shape of the object.

1-7 SECTIONAL VIEWS

Sectional views are used to expose the internal surfaces of an object that would otherwise be hidden from direct view. Figure 1–22 illustrates a comparison between orthographic views and sectional views. Cutting planes lines (see Figure 1–23), are used to define where the

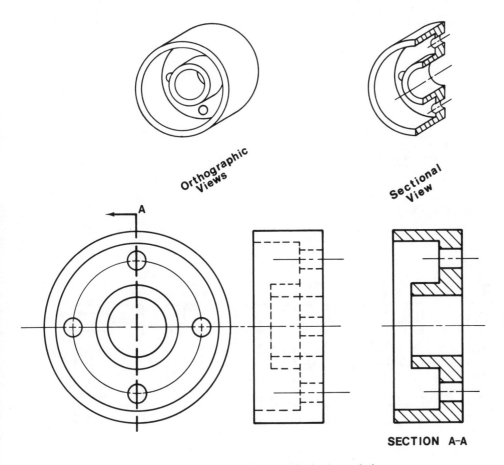

Figure 1-22 An orthographic and sectional view of the same object.

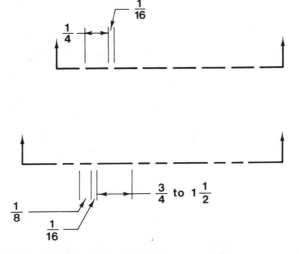

Figure 1-23 How to draw a cutting plane line.

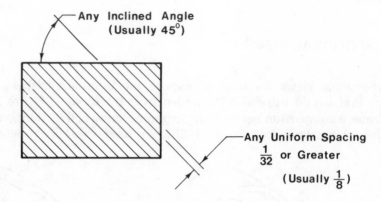

Any Inclined Angle
(Usually 45°)

Any Uniform Spacing
$\frac{1}{32}$ or Greater

(Usually $\frac{1}{8}$)

Figure 1-24   Section lines.

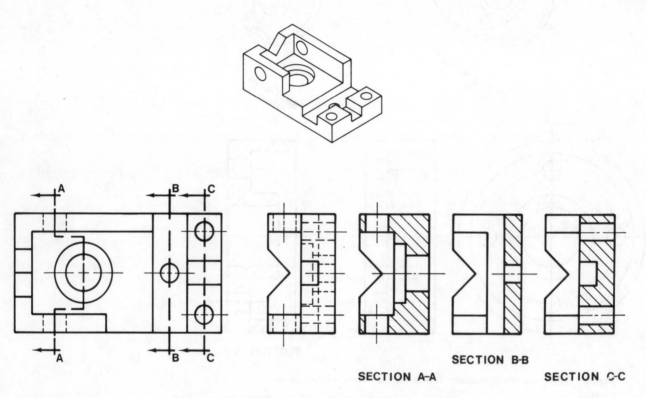

SECTION A-A

SECTION B-B

SECTION C-C

Figure 1-25   Multiple sectional views of the same object.

sectional view is to be taken and section lines (see Figure 1-24), are used to designate those surfaces which have been cut when creating a sectional view. If more than one sectional view is taken on the same object, the sectional views are aligned as shown in Figure 1-25 and identified by letters.

A broken-out sectional view is used when less than a complete sectional view is desired. It is like peeling back part of the outside surface so as to expose part of the internal surfaces. Figure 1-26 illustrates a broken-out sectional view.

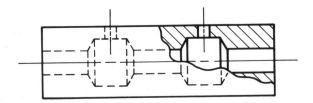

Figure 1-26   A broken out sectional view.

1-8   BASIC DIMENSIONING

Dimensions are used to define the size of an object. They are added to a drawing by using extension, dimension and leader lines as illustrated in Figure 1-8.

All lettering used when dimensioning should be at least $\frac{1}{8}$ (.13) high. Lettering smaller than $\frac{1}{8}$ (.13) is difficult to read, especially on blueprints, and is likely to cause manufacturing errors.

Dimensions may be placed on a drawing using either the unidirectional or aligned systems, although the unidirectional is preferred. Figures 1-27, 1-28, and 1-29 illustrate how to dimension holes, arcs, angles, and small distances. All examples have been dimensioned using the unidirectional system.

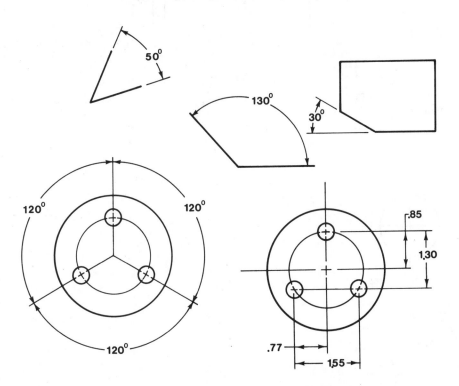

Figure 1-27   Examples of unidirectional dimensioning system.

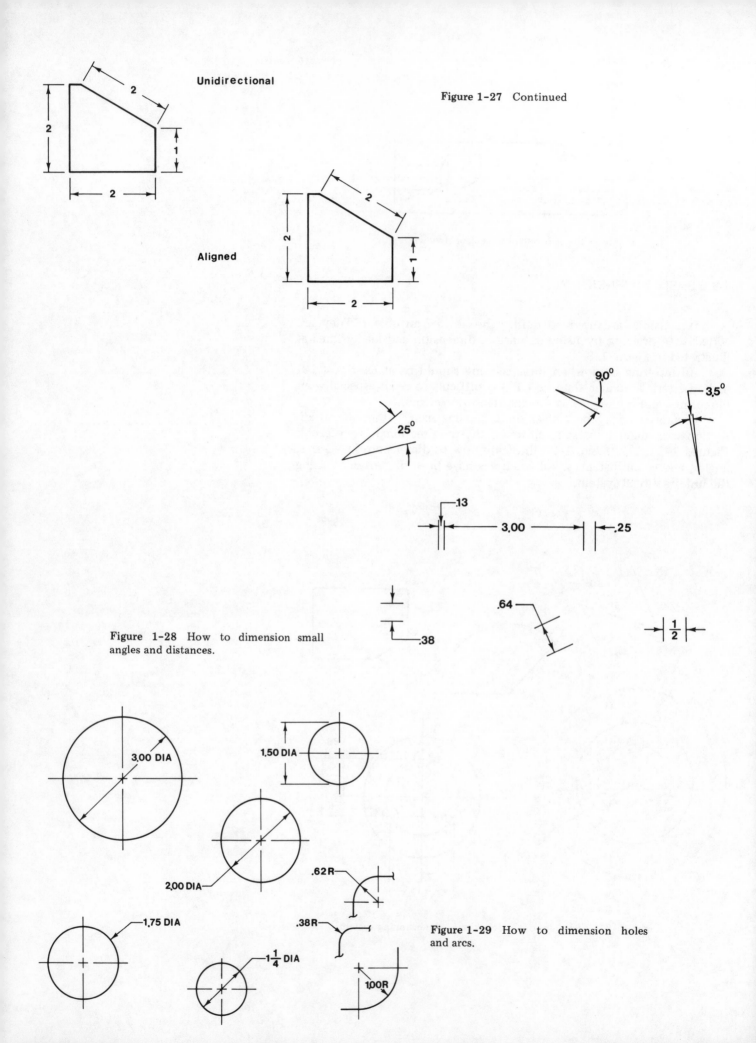

Unidirectional

Aligned

Figure 1-27 Continued

Figure 1-28 How to dimension small angles and distances.

Figure 1-29 How to dimension holes and arcs.

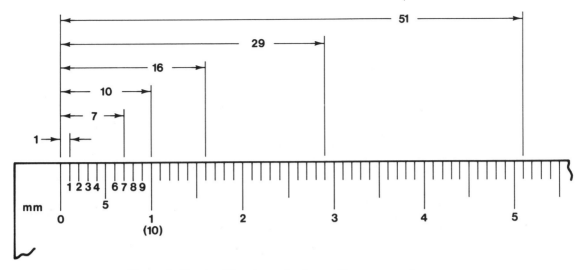

Figure 1-30 A millimeter scale along with some sample measurements.

1-9 METRICS

The metric system is based on a fixed unit of distance, the meter. A meter is divided into smaller units called centimeters or millimeters. There are 100 centimeters or 1000 millimeters to a meter.

The symbol for a millimeter is mm (5 mm, 26 mm, etc.). Figure 1-30 shows a millimeter scale along with a few sample measurements.

For your convenience, conversion tables for both inches to millimeters and millimeters to inches have been included on the inside back cover of this book.

1-10 FASTENERS

There are three different ways to present threaded fasteners on a drawing; the simplified, schematic, and detailed representations. Figure 1-31 shows the three types of representations.

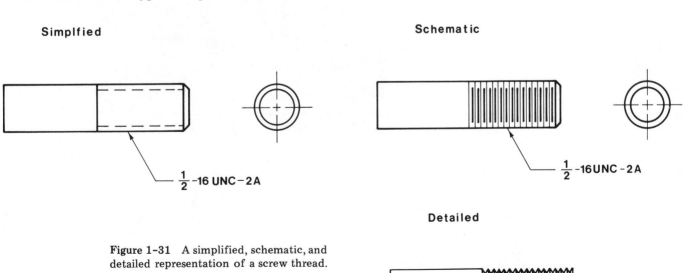

Figure 1-31 A simplified, schematic, and detailed representation of a screw thread.

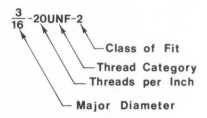

$\frac{3}{16}$-20UNF-2

Class of Fit

Thread Category

Threads per Inch

Major Diameter

Figure 1-32 The meaning of a standard thread notation.

Regardless of which representation is drawn, the same thread notations are used. Figure 1-32 defines the meaning of a standard thread notation.

When drawing threaded holes, always account for the pilot drill hole (a small hole drilled before the threads are cut). When showing a fastener inserted in a threaded hole, remember to clearly draw the bottom edge of the fastener, the remaining unused threaded portion of the hole, and the pilot drill hole as shown in Figure 1-33.

Draftsmen often use fastener templates as guides when drawing fasteners. Figure 1-34 illustrates one of the many fastener templates commercially available, and an Elemoto screw selector which is helpful in picking thread and fastener sizes.

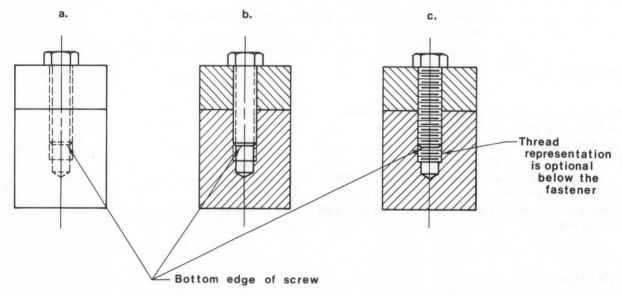

a.   b.   c.

Thread representation is optional below the fastener

Bottom edge of screw

Figure 1-33 How to draw a fastener in a threaded hole.

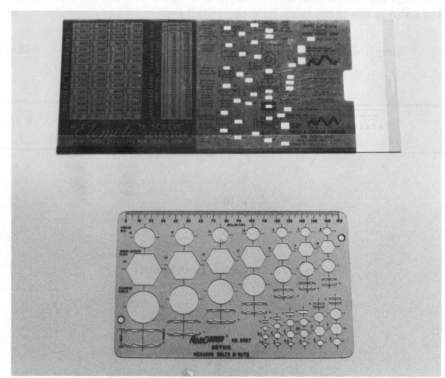

Figure 1-34 A template which is helpful when drawing bolts and nuts, and an ELEMOTO screw selector.

PROBLEMS

1-1 Draw a line which slopes up from left to right at 27° to the horizontal and is 2.50 long. Draw two more lines parallel to the first, one of which is 1.00 above the first and the other 0.75 below.

1-2 Draw a 45° angle and bisect it.

1-3 Draw a line which slopes down from left to right at 30° to the horizontal and is 3.25 long. Divide the line into 5 equal parts.

1-4 Draw an angle of 115° and draw a round of 0.38 within the angle.

1-5 Draw a hexagon 1.00 across the flats.

1-6 Draw a hexagon 0.63 across the flats.

1-7 Draw an ellipse whose minor axis is 1.25 and whose major axis is 2.88.

1-8 and 1-9 Redraw the figure below including the dimensions.

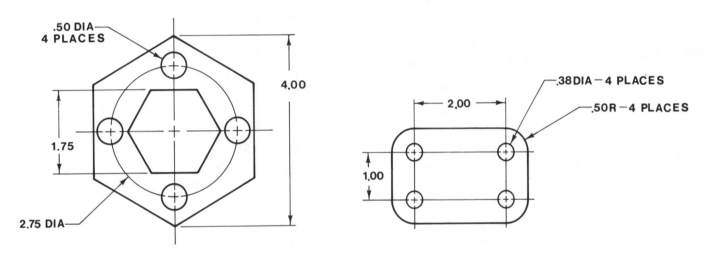

1-10 Redraw the figure below including the dimensions. The dimensions are in millimeters.

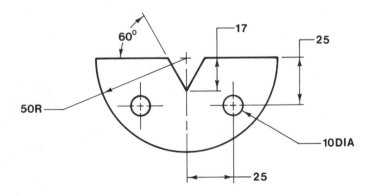

1-11 and 1-12   Redraw the following figures and add all dimensions necessary. Each block on the grid is 0.20 per side.

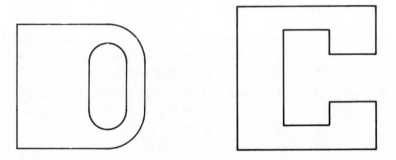

1-13 through 1-19   Draw the front, top, and right side views of the figures below.

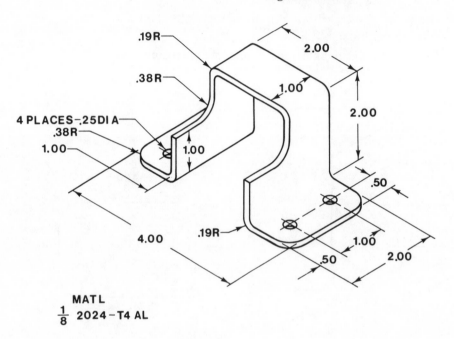

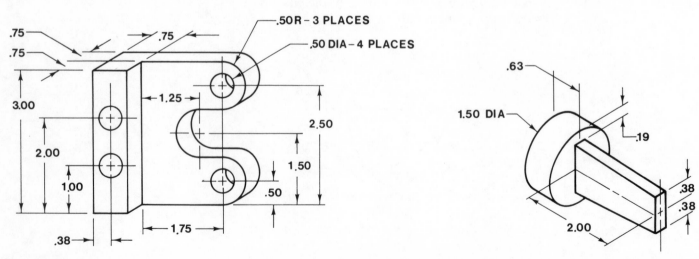

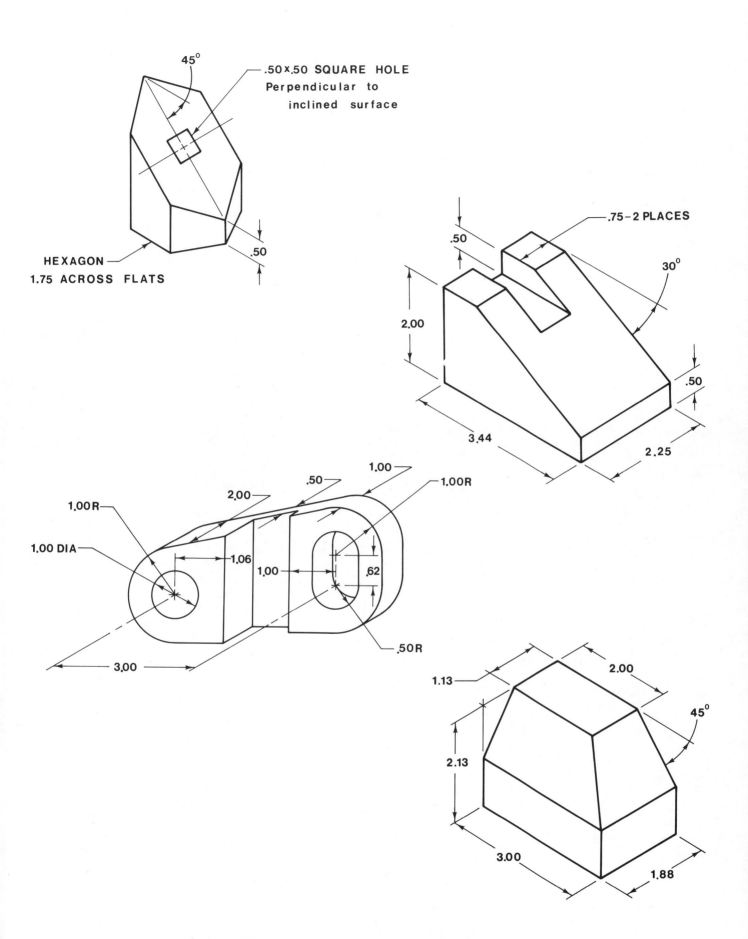

45°

.50×.50 SQUARE HOLE
Perpendicular to
inclined surface

HEXAGON
1.75 ACROSS FLATS

.50

.75-2 PLACES

30°

.50

2.00

3.44

2.25

1.00R

.50

2.00

1.00

1.00R

1.00R

1.00 DIA

1.06

1.00

.62

.50R

3.00

1.13

2.00

45°

2.13

3.00

1.88

21

1-20   Redraw the front and right side views given and
        add the top view.

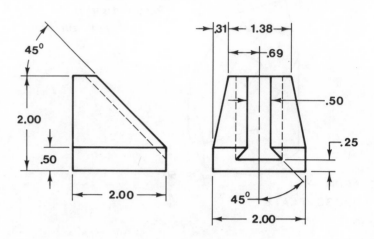

1-21   Redraw the front and top views and add the right
        side view.

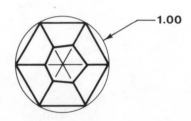

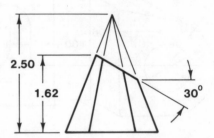

1-22 Redraw the following figure and replace the given front view with a sectional view.

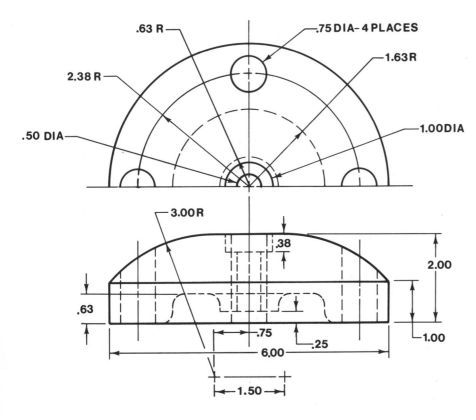

1-23 Redraw the following figure and fasteners.

a. ¼-20 UNC × 1.25 hex head
b. ⅜-24 UNF × 1.50 sq head
c. ½-13 UNC × 1.00 headless stud
d. ⁷⁄₁₆-20 UNF × 1.44 hex head

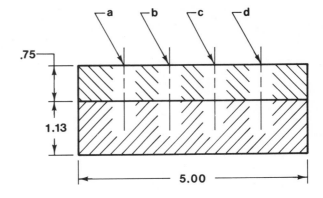

ELECTRONIC SYMBOLS

2-1 INTRODUCTION

This chapter presents some of the many symbols used in electronic drafting to represent components. It also includes a discussion on how to use an electronic symbol template and how to, if needed, create your own symbols.

So that you can better understand the sizes and proportions of each symbol, the symbols have been drawn twice their usual size on a $\frac{1}{8}$-inch grid background. It should be pointed out that the sizes given in this chapter, although in agreement with established national standards, are intended as guides, not as rigorous, absolute measurements. They may be varied slightly as necessary, so long as the proportions of the symbol remain the same.

Consider, for example, the symbol for a resistor shown in Fig. 2–1. The one marked "per standards" is drawn exactly per MIL-15 and ANSI Y32.2 specifications. In addition, two correct and two incorrect

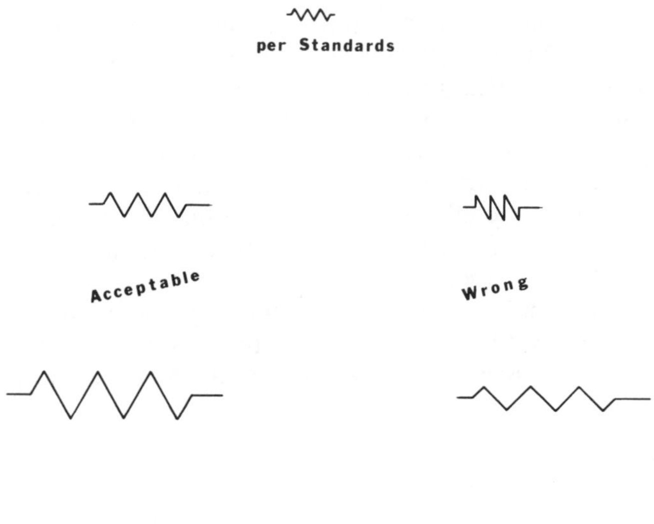

per Standards

Acceptable

Wrong

Figure 2-1 Acceptable and wrong ways to draw electronic symbols.

variations are also shown. In each case, the acceptable variations are correct because they increase the size of the symbol, but do not change the proportions; whereas the unacceptable variations do vary the proportions. When preparing your drawings, choose a scale which best (most clearly) presents the symbols in a neat, easy-to-follow size.

2-2   HOW TO USE A SYMBOL TEMPLATE

Most draftsmen use symbol templates as guides when drawing electronic symbols. They enable the draftsman to work much faster than he would were he using standard drafting equipment (T-square, triangle, compass, etc.) to create the symbols.

There are many different types of symbol templates commercially available; Figure 2-2 shows a few. Some are for general use while others are for more specific purposes such as switches or logic symbols. Regardless of which you choose, they should be of a size which conforms with either ANSI Y32.2 or MIL-15 standards.

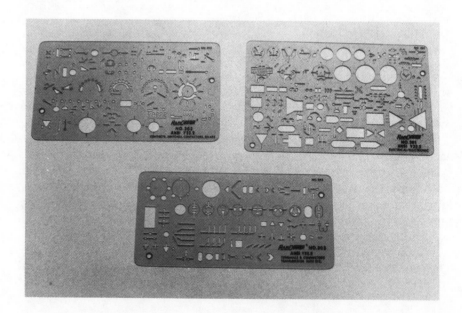

Figure 2-2   Electronic symbol templates.

To use a symbol template, align the appropriate symbol over the place you want it to appear on the drawing and then, holding the pencil as nearly vertical as possible (see Fig. 2-3), trace the symbol using the template as a guide. If a double line appears, it means that the hole cut in the template is too wide. This can be overcome by sliding another template or a triangle under the symbol, as shown in Fig. 2-4. This will raise the template to a thicker portion of the tapered drawing lead. If the lead will not fit into the template, resharpen the lead to a thinner taper so that it will fit.

It should be pointed out that some templates are manufactured in such a way that it is impossible to draw the symbols in one alignment. In these instances, the symbols must be drawn in parts, using several

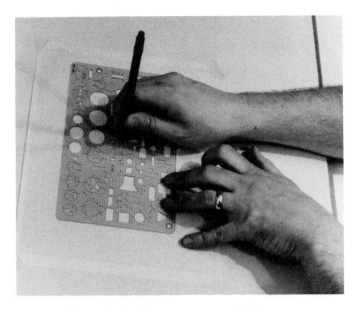

Figure 2-3 How to use a symbol template.

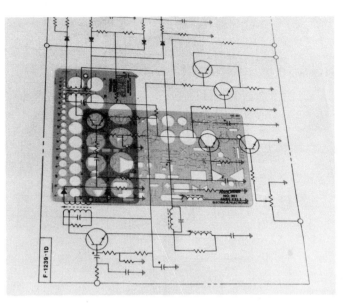

Figure 2-4 If template cutouts are too wide, slide another template underneath and draw as shown in Fig. 2-3.

different cutouts on the template. When this is the case, the final symbol should appear neat and clear, just as though it were drawn in one smooth motion.

2-3 DRAWING PAPER

Most electronic draftsmen prepare their drawings on a grid paper, such as shown in Fig. 2-5. The grid is very helpful in drawing wiring paths and aligning symbols.

The grid is usually printed in nonreproducible blue so that it will not appear on the blueprints made from it. The grid is available in different sizes, $\frac{1}{8}$, $\frac{1}{10}$, $\frac{1}{5}$, $\frac{1}{4}$, etc., but most draftsmen use the $\frac{1}{8}$ size (eight squares to the inch). All the illustrations in this book, drawn on a grid background, were originally drawn on a $\frac{1}{8}$-inch pattern.

Figure 2-5 A sheet of drawing paper with a non-reproducible grid. Squares are 8 to the inch.

Drawing Specifications
(Twice Size)

Per Standards

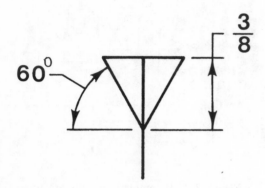

$$\frac{3}{8}$$

60^0

Pictorial   Representation

Related   Symbols

Dipole

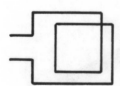

Loop

Counterpoise

**Drawing Specifications
(Twice Size)**   **Per Standards**

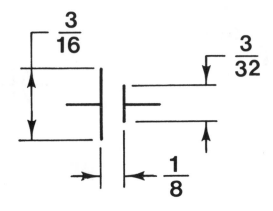

Pictorial   Representation

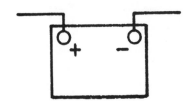

Related   Symbols

Multicell

With Taps

Adjustable

Drawing Specifications
   (Twice Size)

Per Standards

Draw arc by
using a circle
template as a guide.
$\frac{3}{8}$ **DIA** hole

$\frac{3}{16}$

$\frac{1}{16}$

Pictorial   Representation

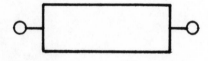

Related   Symbols

Polarized

Shielded

Adjustable

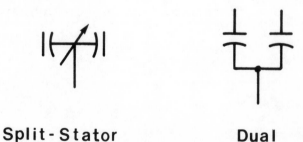

Split - Stator

Dual

32

Drawing Specifications　　　　**Per Standards**
(Twice Size)

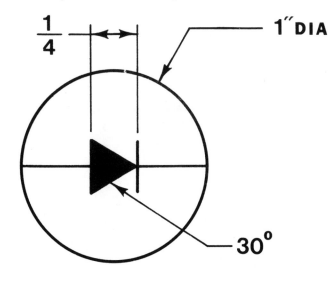

$\dfrac{1}{4}$　　1″ DIA

30°

Pictorial Representation

DIODE

Related Symbols

　　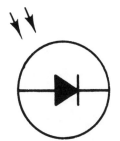　　

Capacitive　　　　**Photosensitive**　　　　**Temperature**
Dependent

Unidirectional　　　　**Tunnel**

Drawing Specifications
(Twice Size)

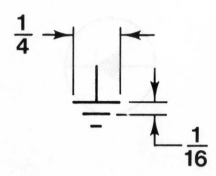

Pictorial Representation

Related Symbols

Chassis

Drawing Specifications (Twice Size)

Per Standards

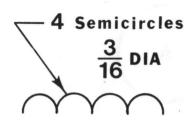

4 Semicircles

$\frac{3}{16}$ DIA

Pictorial Representation

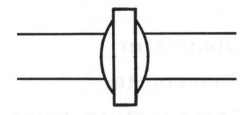

Related Symbols

Magnetic Core

Tapped

Adjustable

Continuous Adjustable

Drawing   Specifications   **Per   Standards**
Pictorial   Representation

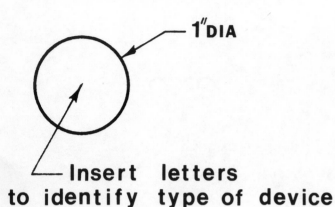

1″DIA

Insert   letters
to identify   type of device

Example

A

LETTER   CODE

A – Ampmeter   OHM – Ohmmeter
CRO – Oscilloscope   PH – Phasemeter
DB – Decibel   Meter   t° – Temperature Meter
F – Frequency   Meter   V – Voltmeter
I – Indicating   W – Wattmeter

Drawing Specifications
(Twice Size)

Per Standards

3 Peaks, both top
and bottom

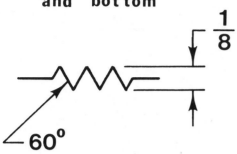

$\frac{1}{8}$

60°

Pictorial   Representation

Related   Symbols

Adjustable   With Taps

Drawing Specifications (Twice Size)

Per Standards

No dimensions
Drawn by eye

Pictorial Representation

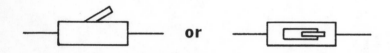

or

Related Symbols

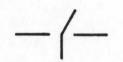

Double-Throw

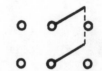

**2 Pole Double
Throw Switch**

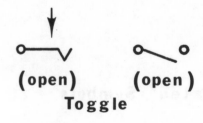

(open)   (open)

Toggle

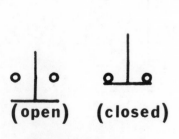

(open)   (closed)

Pushbutton

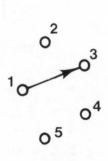

Multiposition

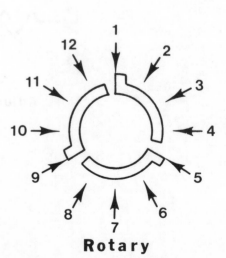

Rotary

Drawing Specifications
(Twice Size)

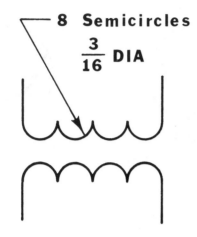

8 Semicircles
$\frac{3}{16}$ DIA

Per Standards

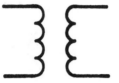

Pictorial Representation

Related Symbols

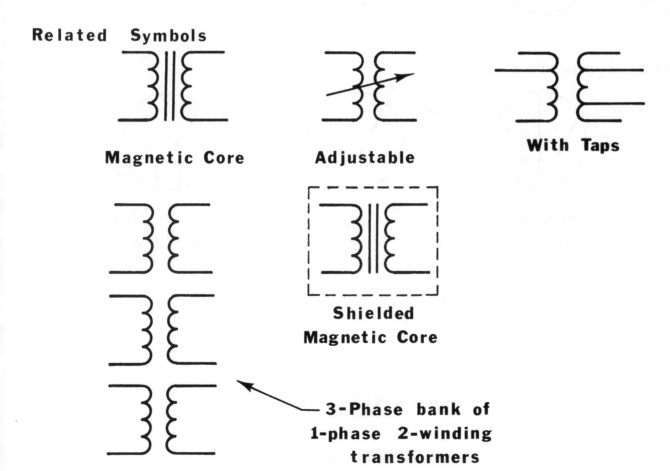

Magnetic Core

Adjustable

With Taps

Shielded
Magnetic Core

3-Phase bank of
1-phase 2-winding
transformers

Drawing Specifications
(Twice Size)

Per Standards

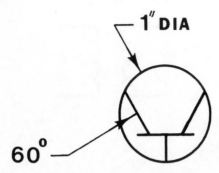

NPN

PNP

Pictorial Representation

Related Symbols

NPN with
Transverse - Biased Base

Unijunction
N - Type

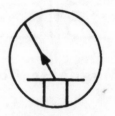

Unijunction
P - Type

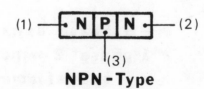

NPN-Type

Amplifier

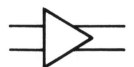

Crystal

Bell

Delay Function

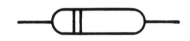

Buzzer

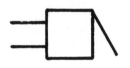

Envelopes

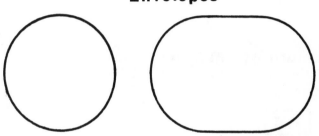

Circuit Breaker

Fuse

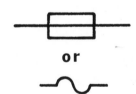

Connectors — Power Supplies

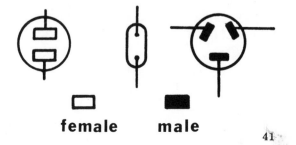

female   male

Headset

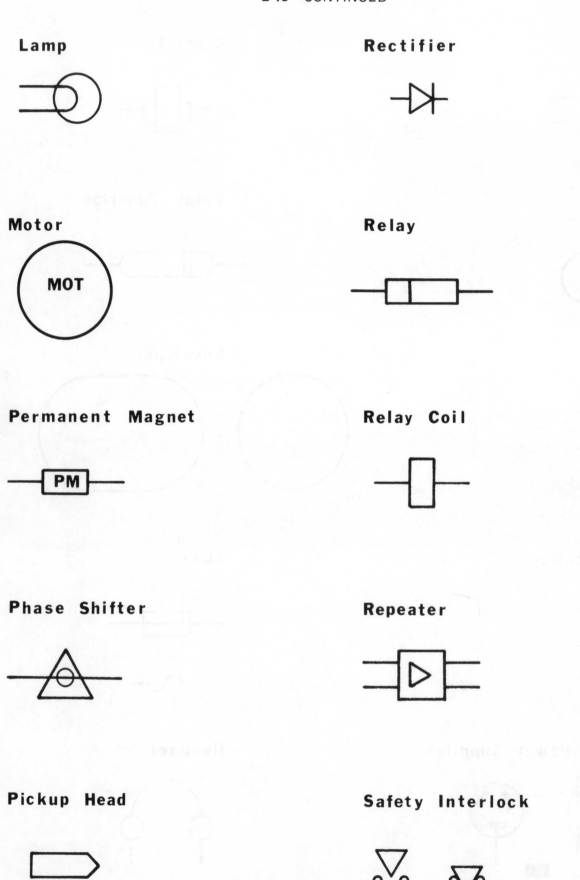

Lamp

Rectifier

Motor

MOT

Relay

Permanent Magnet

PM

Relay Coil

Phase Shifter

Repeater

Pickup Head

Safety Interlock

42

Shielding

Speaker

Terminal   Board

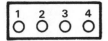

Thermal   Elements

Thermocouple

2-16 SYMBOLS REFERENCES

In addition to the symbols presented in this chapter, there are many more symbols used in electronic drafting. A complete listing of all standard symbols can be found in either of the following publications:

MIL-STD-15-1, Graphic Symbols for Electrical and Electronics Diagram. Washington, D.C.: Dept. of Defense.

USAS Y32.2-1967, Graphic Symbols for Electrical and Electronic Diagrams. New York: United States of America Standards Institute.

Square Reader

Super Light

Transistor X

Figure 2-6   Hypothetical symbols.

2-17 CREATING YOUR OWN SYMBOLS

If, as you are preparing a drawing, you are asked to draw a component for which you are unable to find a predefined symbol, it is acceptable to create your own symbol, provided you completely define the new symbol in the drawing's ledger. It should be emphasized that creating new symbols is ONLY acceptable when you can not find an equivalent in the symbol standards.

When creating your own symbols, any shape or lettering code may be used, although you should be careful not to use shapes or letters which are so similar to well-known symbols that the reader might mistake your new symbol for the known one. Figure 2-6 gives some hypothetical examples of new symbols.

2-18 OTHER WAYS TO DRAW ELECTRONIC SYMBOLS

In addition to using symbol templates as guides, there are several other ways to draw electronic symbols. If the drawing is to be done in ink, a Leroy symbol guide as shown in Fig. 2-7, could be used. Dry

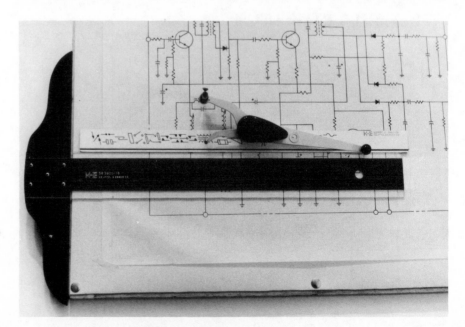

Figure 2-7   Using a Leroy symbol guide to draw electronic symbols.

transfer symbols may be used in conjunction with ink, pencils, or tape drawings. Computers can be programmed to draw not only symbols but also entire drawings including wiring paths and written data. Each of these techniques will produce clear, easy-to-read and easy-to-reproduce drawings, but of the three described, computers offer the best possibilities because of their drawing speed and their designing capabilities.

PROBLEMS

2-1   Using the format outlined in Figure 2–8 and the title block as defined on the inside of the back cover, identify the symbols

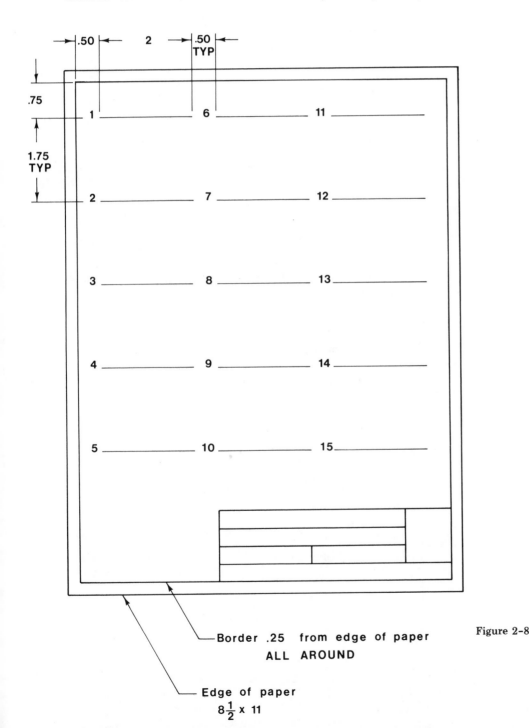

Figure 2–8

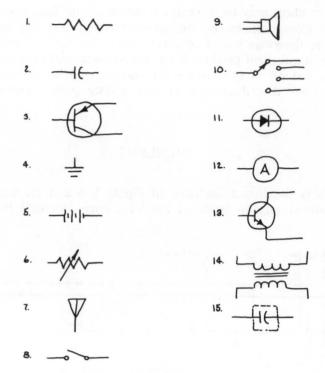

Figure 2-9

sketched in Figure 2-9 and letter the names next to the appropriate numbers.

2-2   Using the format outlined in Figure 2-8, draw and label the following symbols.

    1. Relay coil
    2. Thermocouple
    3. Fuse
    4. PNP transistor
    5. Dipole antenna
    6. Temperature-dependent diode
    7. Polarized capacitor
    8. Adjustable resistor
    9. Chassis ground
  10. Phase shifter
  11. NPN transistor
  12. Tapped inductor coil
  13. Battery
  14. Bell
  15. Circuit breaker

2-3   Using the format outlined in Figure 2-8, draw and label the following symbols.

1. Delay function
2. Loop antenna
3. Amplifier
4. Unijunction P-type transistor
5. Tapped resistor
6. Ampmeter
7. Multicell battery
8. Safety interlock
9. Counterpoise antenna
10. Ground
11. Split stator
12. Buzzer
13. Permanent magnet
14. Resistor
15. Ohmmeter

2-4   Using the format outlined in Figure 2-8, draw and label the following symbols.

1. Photosensitive diode
2. Shielded capacitor
3. Antenna
4. Continuous adjustable inductor
5. Terminal board
6. Toggle switch
7. Transformer
8. Speaker
9. Motor
10. Thermal element
11. Temperature-dependent diode
12. Voltmeter
13. Pickup head
14. Pushbutton
15. Dipole antenna

2-5   Draw and label the following pictorial symbols.

1. Antenna
2. Transistor
3. Inductor
4. Resistor
5. Capacitor
6. Battery

SCHEMATIC DIAGRAMS

3-1 INTRODUCTION

Schematic diagrams are drawings which show graphically what components are to be used and how these components are to be connected to form a desired circuit. There are two basic types of schematic diagrams: those which use electronic symbols and those which use pictorial symbols. This chapter will present both types of diagrams.

3-2 HOW TO DRAW SCHEMATIC DIAGRAMS

When drawing schematic diagrams, it is important that the drawing be easy to follow. The symbols should be drawn clearly and neatly with heavy, black lines. All relative data, such as component values and identification numbers, should be located as close as possible to the appropriate symbols [Figure 3-1(a)].

To lay out the diagram clearly, the individual symbols should not be cramped together. A crowded group of symbols is difficult to read and is therefore more likely to cause error. Cramping can be avoided by planning the placement of the symbols ahead of time.

One method which may be used to help assure neat, uncluttered schematic diagrams is to divide the original design sketches into quarters as shown in Figure 3-2. These quarters need not be equal in area, but may be varied so as to include approximately a fourth of the symbols. This will give you an idea as to what quarter of the final drawing the symbols should be included in.

After quartering the design sketches, lay out *equal* quarters on the drawing paper, then draw the components which will take up the most space on the diagram (rotary switches and transistors, for example). By

51

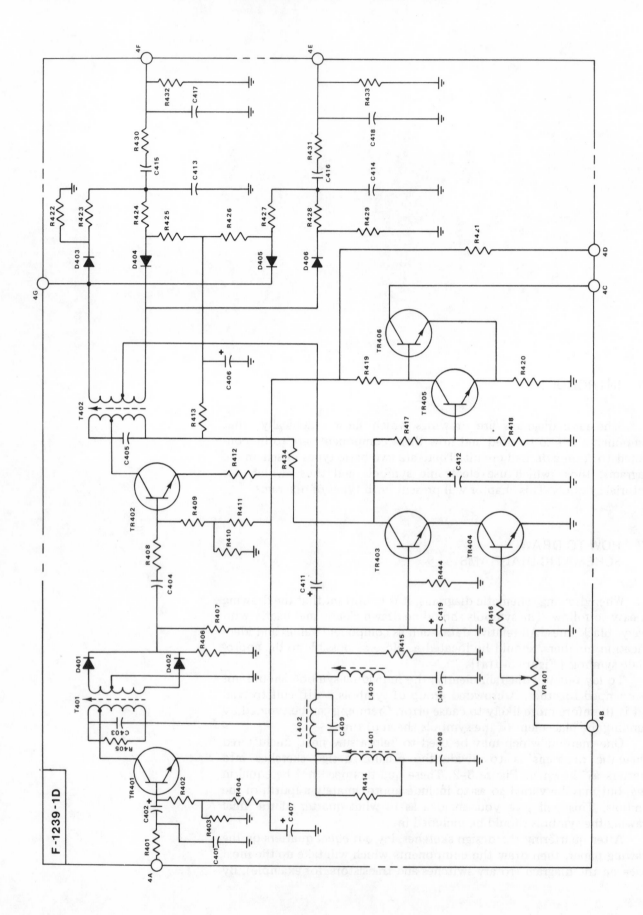

Figure 3-1(a)

F-1239-1D

Part No	Value / Name	Part No	Value / Name
R401	1 k	C412	1 UF 50V EL
R402	100 k	C413	680 PF ⎫ ±5% 50V ST
R403	15 k	C414	680 PF ⎭
R404	22 k	C415	0.15 UF ⎫ ±10% 50V MY
R405	68 k	C416	0.15 UF ⎭
R406	100 k	C417	0.006 UF ⎫ ±5% 50V MY
R407	100 k	C418	0.006 UF ⎭
R408	4.7 k	C419	1 UF 50 EL
R409	100 k		
R410	2.2 k		
R411	22 k		
R412	33 k	TR401	2SC711
R413	220 k	TR402	2SC711
R414	47 k	TR403	2SC711
R415	2.2 k ALL VALUES ARE	TR404	2SC711
R416	47 k IN OHMS	TR405	2SC733
R417	22 k ±10% $\frac{1}{4}$W CB	TR406	2SC735
R418	22 k		
R419	3.3 k		
R420	4.7	D401	IN34A
R421	47	D402	IN34A
R422	220 k	D403	IN34A
R423	10 k	D404	IN34A
R424	10 k	D405	IN34A
R425	220 k	D406	IN34A
R426	220 k		
R427	10 k		
R428	10 k		
R429	220 k	T401	19 kHz Coil
R430	56 k	T402	38 kHz Coil
R431	56 k		
R432	15 k		
R433	15 k	L401	19 kHz Coil
R434	47 k	L402	Micro Inductor
		L403	67 kHz Coil
C401	68 PF ±10% 50V CE		
C402	10 UF 10V EL		
C403	10000 PF ± 5% 50V ST		
C404	0.022 UF ±10% 50V MY		
C405	4700 PF ± 5% 50V ST		
C406	1 UF 50 V EL		
C407	47 UF 25 V EL		
C408	10000 PF ⎫		
C409	2200 PF ⎬ ± 5% 50V ST		
C410	270 PF ⎭		
C411	10 UF 25V EL		

Figure 3-1   (a) A schematic diagram for the FM MPX block of a Sansui 350A Solid-State AM/FM Stereo Tuner Amplifier. The diagram is courtesy of Sansui Electric Co., Japan; (b) a schematic diagram parts list for the Sansui 350A shown in Fig. 3-1(a).

first assuring that these larger symbols are spaced far enough apart so that the diagram does not appear cluttered, the smaller symbols (such as resistors and capacitors) can easily be added.

Component values and/or stock numbers may be included by lettering the data next to the component as shown in Figure 3-3 or by setting up a table as shown in Figure 3-1(b). The table method has the advantage of allowing purchasers to work from a specific list of components rather than creating their own lists from the schematic.

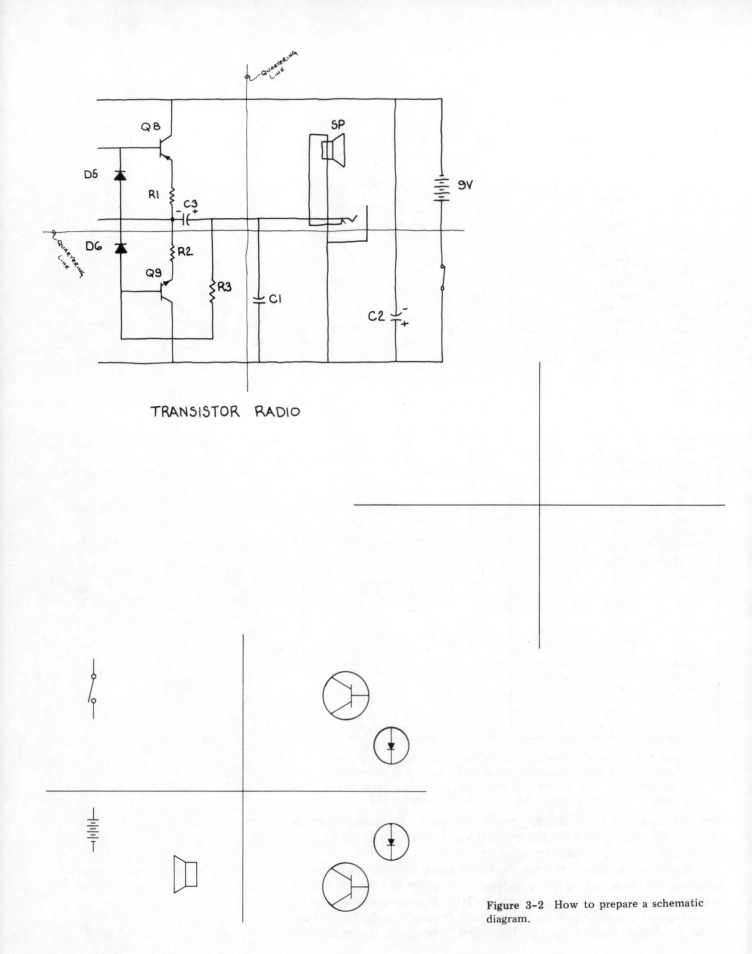

TRANSISTOR RADIO

Figure 3-2 How to prepare a schematic diagram.

54

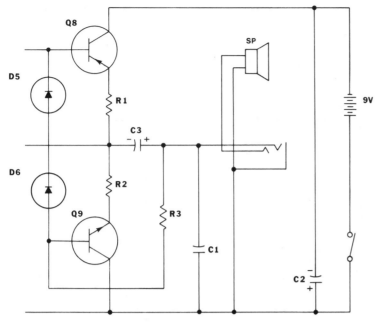

Figure 3-2 Continued.

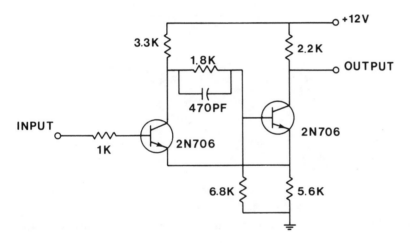

SCHMITT TRIGGER

Figure 3-3 A schematic diagram.

Including the data next to the symbols allows technicians to more quickly read the diagram without constantly referring back to the table.

If possible, arrange the schematic diagram so it can be read from left to right in terms of function. Place symbols which work together to perform a function together on the drawing.

3-3 CROSSOVERS AND INTERSECTIONS

There are two different conventions used to represent crossovers and intersections. The older convention, which is now rarely used, pictures crossovers as loops (semicircles) and intersections as two lines crossing. Figure 3-4 illustrates both types. The newer convention is the dot convention which indicates intersections with dots and crossovers

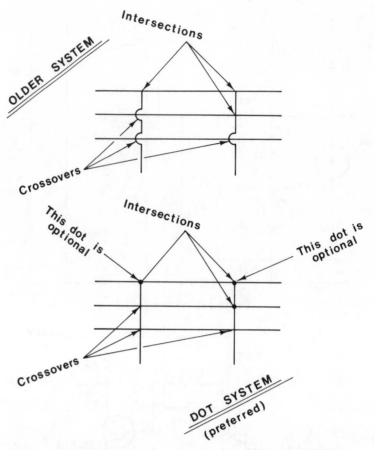

Figure 3-4 How to draw crossovers and intersections on a schematic diagram.

by crossed lines (see Figure 3-4). With the dot system, when a line joins a line at 90° to form an obvious intersection, the dot may be omitted.

3-4 NOTATIONS AND VALUES

Each numerical value given on a schematic diagram must include the units of measure. For example, capacitors are measured in farads, so each capacitor value must be accompanied by the farad unit of measure: i.e., 2 farads, 0.4 farads.

Each unit of measure has a symbol to make it faster to print on a drawing. Figure 3-5 lists the units of measure and their symbols for components most often called for on schematic diagrams.

Units of measure often include a multiplier. A kilowatt for example, means 1000 watts or 1×10^3. Kilo is a multiplier which when used as a prefix to a unit of measure means the unit of measure is to be multiplied by 1000. A megaton means 1,000,000 tons. Mega means 1,000,000 or 1×10^6. Figure 3-6 shows additional examples.

COMPONENT	UNITS	SYMBOL
Resistor	Ohms	Ω
Capacitor	Farads	F
Inductor	Henrys	H
	Watts	W
	Volts	V
	Alternating Current	AC
	Direct Current	DC

Figure 3-5 Notations and values.

PREFIXES

Figure 3-6 Prefixes for multipliers and submultipliers.

Prefix	Symbol	Multiples and Submultiples
Tera	T	10^{12}
Giga	G	10^{9}
Mega	M	10^{6}
Kilo	K	10^{3}
Hecto	H	10^{2}
Deka	DA	10
Deci	D	10^{-1}
Centi	C	10^{-2}
Milli	m	10^{-3}
Micro	U	10^{-6}
Nano	N	10^{-9}
Pico	P	10^{-12}
Femto	F	10^{-15}
Atto	A	10^{-18}

Examples

2 Picofarads = 2PF = .000 000 000 002 F

4 Milliohms = 4mΩ = .004 Ω

6 Kilowatts = 6KW = 6000W

A submultiplier is a multiplier which has a negative exponent. A microfarad is equal to 0.001 farads. A micro means to multiply by 1/1000 or 1×10^{-3}. A pico means 1/1,000,000,000,000 or 0.000 000 000 001 or 1×10^{-12}. We can see how much easier the symbols for prefixes make writing values and units of measure. A picofarad would be written on a drawing as: 1 PF.

It is sometimes difficult to keep track of all the decimal places involved in using multipliers. How many decimal places (zeros) are added to 23.6 when the number is written 23.6×10^5? For positive exponents (the small numbers written above and to the right of the 10's), add decimal places to the right of the decimal point equal to the exponential value. Therefore 23.6×10^5 equals 2360000. For negative exponents, add decimal places to the left. 23.6×10^{-5} equals 0.000236. Figure 3-7 illustrates this method.

$$23.6 \times 10^4 \quad = \quad 236,000$$

$$23.6 \times 10^3 \quad = \quad 23,600$$

$$23.6 \times 10^2 \quad = \quad 2,360$$

$$23.6 \times 10^1 \quad = \quad 236$$

$$23.6 \quad = \quad 23.6$$

$$23.6 \times 10^{1} \quad = \quad 2.36$$

$$23.6 \times 10^{2} \quad = \quad .236$$

$$23.6 \times 10^{3} \quad = \quad .0236$$

$$23.6 \times 10^{4} \quad = \quad .00236$$

Figure 3-7 How to read multiples of 10.

3-5   BASIC CIRCUITS

Figure 3-8 illustrates some of the more common circuit configurations. These basic configurations are found as part of many more sophisticated circuit designs. The draftsman should be familiar enough with these basic circuits to recognize them when they appear in design sketches.

3-6   PICTORIAL SCHEMATIC DIAGRAMS

Pictorial schematic diagrams are schematic diagrams which use the pictorial symbols defined in Chapter 2 instead of the national standard symbols. These schematics are useful to persons not familiar with the standard symbols, but they lack the explicit detail required by designers. Figure 3-9 shows a schematic diagram and the same diagram in pictorial form.

Pictorial schematics are also helpful in the preparation of printed circuit drawings, which are covered in Chapter 7.

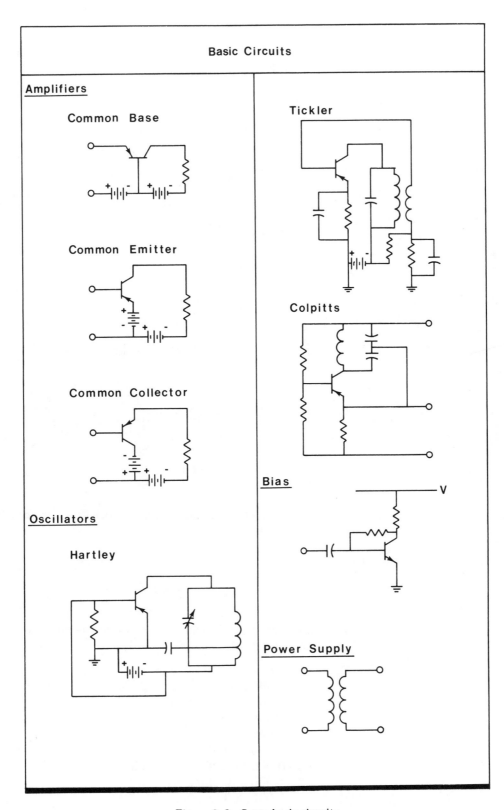

Figure 3-8 Some basic circuits.

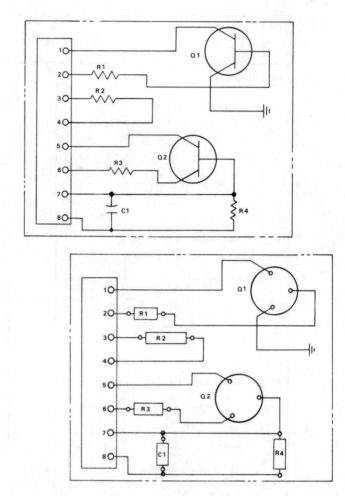

Figure 3-9 A schematic diagram and the same circuit as a pictorial schematic diagram.

PROBLEMS

3-1 Express the following values in words and then as they would be printed on a drawing.

1. 0.003 amps
2. 0.000 000 004 farads
3. 1000 ohms
4. 10,000 watts
5. 2,500,000 ohms
6. 0.000 000 000 025 farads
7. 4,000,000 ohms
8. 0.000 003 farads
9. 0.025 amps
10. 1500 volts

3-2 Write the numerical equivalents of the following. Be sure to include the units of measure.

1. 4 PF
2. 3 KV
3. 1.5 mA
4. 2 KΩ
5. 6 KW
6. 1.5 Ω
7. 3 NF
8. 1.75 MΩ
9. 2 UA
10. 1 MW

3-3 Redraw the Colpitts oscillator circuit shown in Figure 3-10 and replace the letters R1, R2, R3, C1, C2, L1, and Q1 with the appropriate symbol and value. The letters are defined as:

Letter	Component	Value
R1	Resistor	12 K
R2	Resistor	8.2 K
R3	Resistor	1.5 K
C1	Capacitor	0.10
C2	Capacitor	0.047
L1	Inductor	10 MH
Q1	Transistor	2N2926

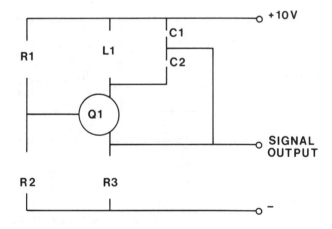

COLPITTS OSCILLATOR

Figure 3-10

3-4 Redraw the basic logic circuit shown in Figure 3-11 and add the following values.

R1   10 KΩ
R2   4.7 KΩ
Q1   2N708   (NPN transistor)

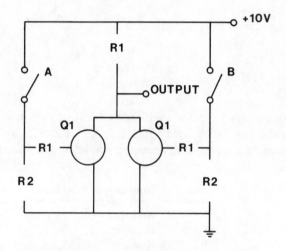

BASIC LOGIC CIRCUIT

Circuit courtesy of General Electric Co.

Figure 3-11

3-5 Redraw the power supply circuit shown in Figure 3-12.

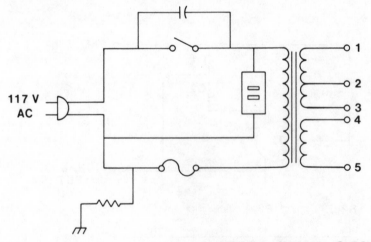

POWER SUPPLY CIRCUIT

Circuit courtesy of General Electric Co.

Figure 3-12

3-6 Redraw the AM broadcast band transmitter shown in Figure 3-13 and substitute as follows.

1. 670 KC   CRYSTAL
2. 150 KΩ
3. 2N170   (NPN transistor)
4. 300 PF
5. Ground
6. Ground
7. 2N170   (NPN transistor)
8. 0.002 PF
9. Ground
10. Inductor   (10 MHY RFC)
11. 1 KΩ
12. Battery 6V.
13. Ground
14. 0.001 PF
15. 0.003 PF
16. L3
17. Ground
18. 10 mF, 10 V
19. 0.001 PF
20. Antenna
21. Chassis ground

Add both symbols and values.

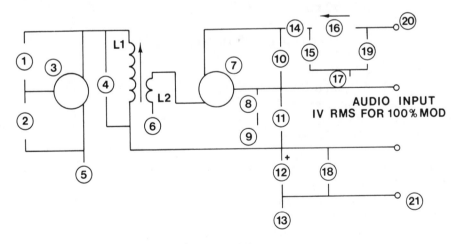

AM BROADCAST BAND TRANSMITTER

Figure 3-13

3-7   Redraw the schematic diagram for the 6-volt phono amplifier shown in Figure 3–14 and substitute as follows.

1. Letter in the words CRYSTAL CARTRIDGE
2. 10 KΩ
3. 8 mF
4. 2N323   (transistor)
5. 150 KΩ
6. 0.5 mF
7. 6.8 KΩ
8. R1   (tapped resistor)
9. 1 KΩ
10. 0.2 mF
11. 0.02 mF
12. 10 KΩ
13. 0.05 mF
14. R3   (tapped resistor)
15. R2   (tapped resistor)
16. 100 KΩ
17. 0.1 mF
18. 220 KΩ
19. 2N323   (transistor)
20. 6 V, 10 mF
21. 2.2 KΩ
22. 10 mF
23. 4.7 KΩ
24. 2N323   (transistor)
25. 47 KΩ
26. 1.5 KΩ
27. 330 Ω
28. 6 V, 50 mF
29. 33 KΩ
30. T1   (iron core transformer)
31. 220 Ω
32. 6 V, 50 mF
33. 6 V, 50 mF
34. Ground symbol
35. 1.2 KΩ
36. 2N1415   (transistor)
37. 33 Ω
38. 2n1415   (transistor)
39. Ground symbol
40. Single throw switch
41. 6 V battery
42. T2   (iron core transformer)
43. Letter in the words TO SPEAKER
44. Ground symbol

All transistors are PNP types. In addition, add the following notes and performance data.

NOTES:

R1 - Bass Control - 50 K Linear Taper
R2 - Treble Control - 50 K Linear Taper
R3 - Volume Control - 10 K Audio Taper
T1 - Driver Transformer - PR1 2K/sec 1.5 K, C.T.
T2 - Output Transformer - PR1 100 Ω/sec V.C. (3.2, 8, 16 Ω)
All resistors ½ watt

PERFORMANCE DATA

Max Power Out at 10% Dist	300 mw
Distortion at 100 mw	60 cycles - 3%
	1.0 KC - 1.5%
	5.0 KC - 3.0%

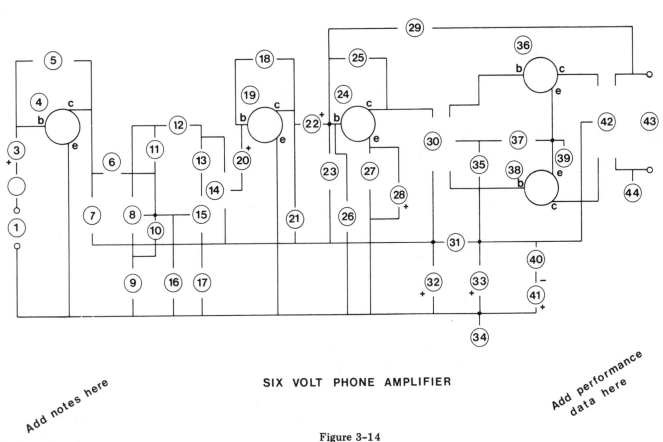

SIX VOLT PHONE AMPLIFIER

Add notes here

Add performance data here

Figure 3-14

3-8   Much of the work assigned beginning draftspersons is in the form of freehand sketches. An engineer will present freehand design sketches and have the draftsperson prepare the finished schematic drawings from these sketches.

Study the following design sketches and redraw them using drawing instruments.

a. Fig 3-15
b. Fig 3-16
c. Fig 3-17

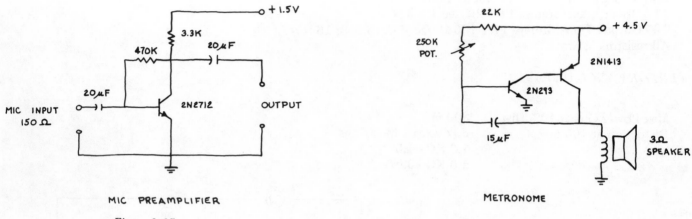

MIC PREAMPLIFIER

Figure 3-15

METRONOME

Figure 3-16

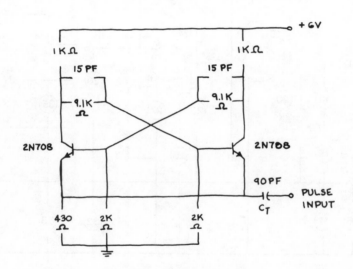

EMITTER TRIGGERING

Figure 3-17

3-9   Redraw the Colpitts circuit of Fig 3-10 as a pictorial schematic.

3-10 Redraw the MIC preamplifier circuit of Fig 3-15 as a pictorial schematic.

CONNECTION DIAGRAMS

4-1 INTRODUCTION

This chapter explains how to draw four different types of connection diagrams. Connection diagrams are drawings which define how various components of a system are to be wired together. They are used most often in conjunction with assembly or maintenance instructions and as design layouts.

Each of the four types of connection diagrams discussed has advantages and disadvantages in its preparation and its use. As you study how to draw the diagrams, try to become aware of these advantages and disadvantages so that you can learn which type is best suited for the requirements of each drawing.

4-2 POINT-TO-POINT DIAGRAMS

Point-to-point diagrams are diagrams which show the terminal connection location and routing path of every wire used in a system. Wires are not shown in bundles nor do they use destination codes or tables. Each line drawn represents one wire which is graphically shown starting at one point and ending at another.

Point-to-point diagrams are very useful in design work because they enable the reader to directly follow the path of each wire. However, for large drawings, which contain many wires, point-to-point diagrams can become confusing and difficult to read accurately.

To draw a point-to-point diagram (see Figure 4–1):

1. Draw the components, locating them in the same position that they are located in the installation. Label all components and terminals.

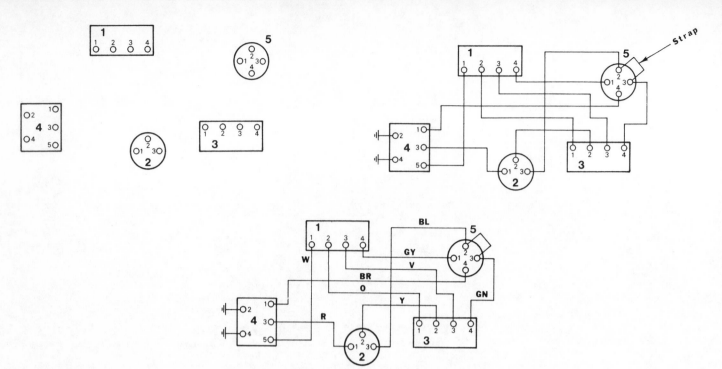

Figure 4-1 How to prepare a point-to-point diagram.

2. Draw in the wire paths. Wires which connect terminals on the same component are called straps.

3. Assign, if required, colors to each wire. A table of wire color codes is included in the appendix.

Some draftsmen prefer to rearrange the terminal numbers so that they can make neater line patterns which are easier to follow. Figure 4-2 illustrates this technique. Note how, by changing the sequence of the terminal numbers, the diagram becomes much neater in appearance and much easier to follow.

4-3   BASELINE DIAGRAMS

Baseline diagrams are diagrams which feed all wires into one central line called a baseline. They present the wiring in a well-organized, easy-to-follow format, but have a drawback in that they do not show components in their correct physical positions. Baseline drawings are

Figure 4-2 An example of how rearranging the terminal numbers can be used to clarify a point-to-point diagram.

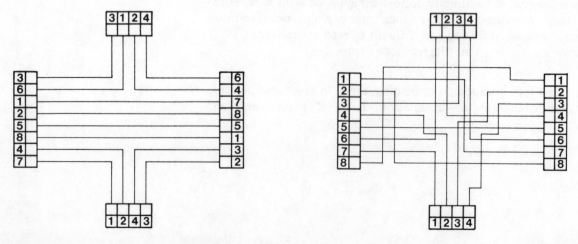

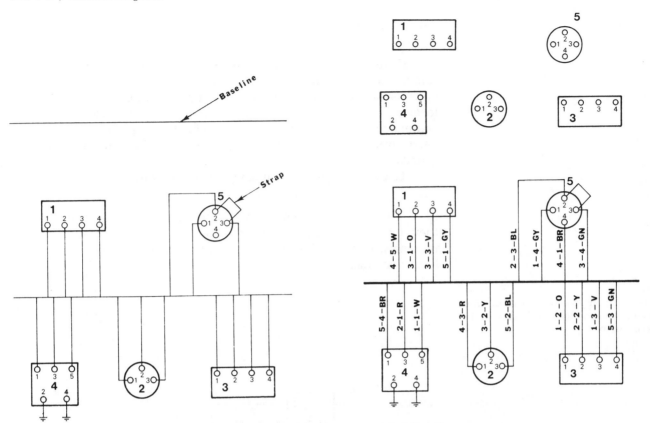

Figure 4-3   How to prepare a baseline diagram.

particularly useful in presenting large, many-component diagrams, in a size suitable for book use. They are used most often for maintenance and assembly manuals.

To draw a baseline diagram (Figure 4-3 illustrates):

1. Draw a light horizontal line across the center of the paper. This line is called the baseline.

2. Draw half of the components above the baseline and half below. Label all components and terminals.

3. Connect all used terminals to the baseline by drawing lines directly from the terminals to the baseline. If a direct path is not possible, make all directional changes at 90°. The baseline may not be bypassed. All wires must be connected to the baseline. (Straps need not be connected to the baseline.)

4. Label all wires using a destination code.

The destination code consists of letters and numbers which identify the component number and terminal number to which the wire is to be attached, and the color of the wire. See Figure 4-4 for a further explanation of the destination code.

5. Darken in the baseline making it a very heavy, black line.

Destination Code

Component No – Terminal No – Wire Color

Example

4 – 8 – W

MEANS a white wire goes to component 4, terminal 8.

Figure 4-4   Destination code.

Highway diagrams are diagrams which combine groups of wires running along similar paths into bundles called highways. The components are located in the same relative positions as they would be located in the actual component setup. Of the different types of component diagrams shown in this chapter, highway diagrams most nearly duplicate the wiring configurations as they appear in the final installation.

To draw a highway diagram (Figure 4-6 illustrates):

1. Draw the components, locating them in the same positions they occupy in the final installation. Identify all components and terminals. Figure 4-5 is a point-to-point diagram we wish to redraw as a highway diagram.

Figure 4-5 A point-to-point diagram which is to be redrawn as a highway diagram and as a lineless diagram.

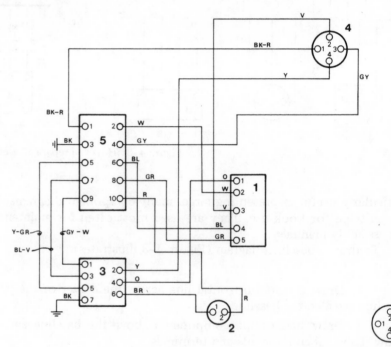

Figure 4-6 How to prepare a highway diagram.

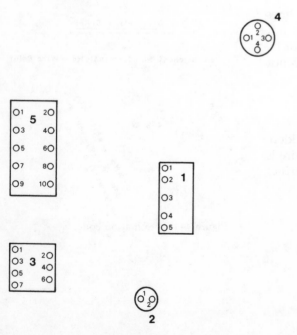

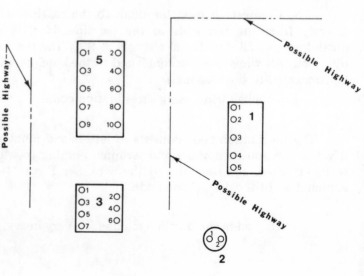

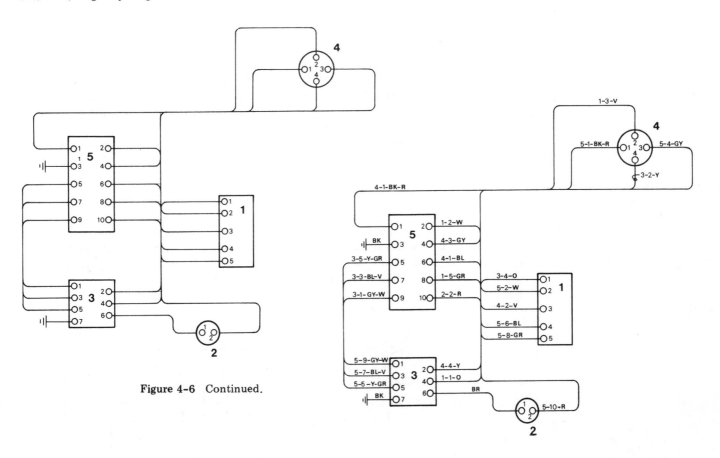

Figure 4-6 Continued.

2. Study the drawing and with light construction lines, identify all possible areas where wires may be bundled. These bundles are called highways.

3. Draw very light lines from the appropriate terminals to the highways. Where the individual wire joins the highway, use either an arc or a line slanted at 45° as shown in Fig. 4-7. The direction of the arc or slanted line should be in the direction the wire is headed. For example, the arc showing the intersection of the wire from terminal 10 of component 5 to the highway turns down as the wire is headed for component 2, whereas the wire from terminal 3 of component 1 turns up as the wire is headed for component 4.

Note that unlike baseline diagrams, highways may be bypassed. The wire from terminal 6 of component 3, for example, may be drawn directly to terminal 1 of component 2. It need not be included as part of a highway.

4. Darken in all lines and label each wire with a destination code; component number—terminal number—wire color. Ground wires need only be labeled with their color and symbol for ground.

Figure 4-7 Wire turns and intersections on a highway drawing may be drawn as either an arc or as a line slanted at 45°.

or

45°   ¼ DIA

Lineless diagrams are component diagrams which do not pictorially show any wiring paths but instead define the wiring paths by using a wiring table. Lineless diagrams are particularly useful for clarifying large complex drawings where the number of wires is so great that several sheets of drawings would be required to include the entire pattern. The disadvantage of lineless diagrams is that because they omit all line patterns, they present an incomplete representation of what the final wiring setup will look like.

To draw a lineless diagram (Figure 4-8 illustrates):

1. Draw the components of the system. If possible, locate the components on the drawing in positions that approximate their actual physical locations, and number the terminals. Each component must be labeled. Figure 4-5 is a point-to-point diagram we wish to redraw as a lineless diagram.

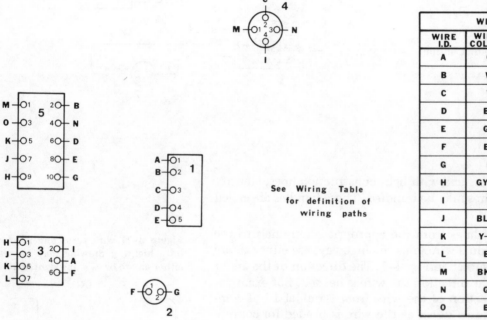

Figure 4-8   An example of a lineless diagram.

2. Prepare a wiring chart such as the one pictured in Figure 4-9. The chart should include wire identification, where the wire starts (component and terminal), and where it ends (component and terminal). The chart may also include wire size and wire color. In this example, the wires were identified by letters and the components by numbers. Component names could also have been used.

3. Draw a short line from each terminal used and label it with the appropriate wire identification letter.

74

WIRING TABLE			
WIRE I.D.	WIRE COLOR	FROM	TO
A	G	1–1	3–2

Sample Callout

Figure 4-9 An example of a wiring table that would accompany a lineless diagram.

PROBLEMS

4-1 Given the point-to-point diagram in Figure 4-10, redraw the diagram as:

    a. highway diagram
    b. baseline diagram
    c. lineless diagram

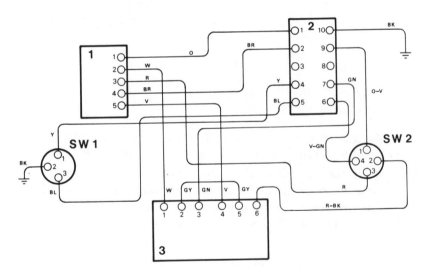

Figure 4-10

4-2 Figure 4-11 is an illustration of a home stereo system which includes two speakers, an amplifier, and a turntable. Redraw the system as:

    a. highway diagram
    b. baseline diagram
    c. lineless diagram
    d. point-to-point diagram

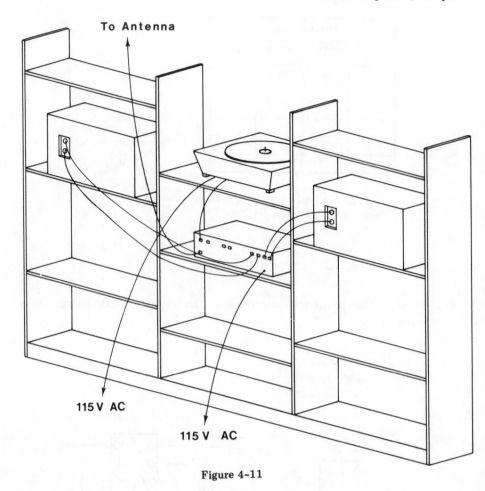

To Antenna

115 V AC

115 V   AC

Figure 4-11

Assign component numbers, terminal numbers, and wire colors as needed. A table of wire color abbreviations may be found.

4-3   Your company is working on a new electronic system. A prototype has been set up in the lab and you have been asked to study the prototype setup (pictured in Figure 4-12) and prepare a wiring diagram. Redraw the system as:

   a. highway diagram
   b. baseline diagram
   c. lineless diagram
   d. point-to-point diagram

4-4   Figure 4-13 is an engineer's design sketch. Redraw the sketch as:

   a. highway diagram
   b. baseline diagram
   c. lineless diagram

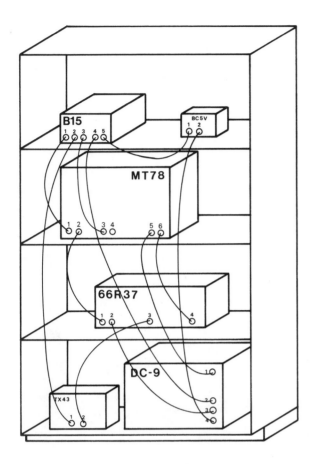

Figure 4-12

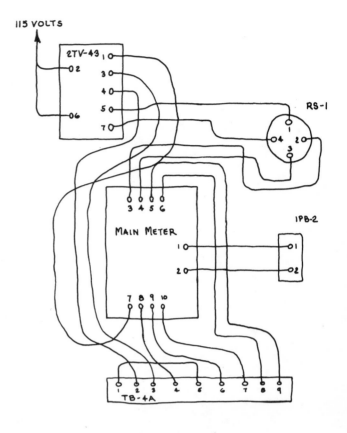

Design Sketch - Diplex System

Figure 4-13

BLOCK AND LOGIC DIAGRAMS

5-1 INTRODUCTION

This chapter explains how to draw and interpret block and logic diagrams. As in previous chapters, the drawing fundamentals required are explained in detail, but in addition, a discussion of the usage of each type of diagram is also included.

5-2 BLOCK DIAGRAMS

Block diagrams are a way to graphically express the relationship between a series of elements. They are used in the electronic and electrical fields, and in almost every other technical area of study: business, mathematics, engineering, to name just a few. Figure 5-1 is an example of a block diagram which illustrates a stereo system.

To draw a block diagram, no special block size or shape is required although rectangular blocks are generally used. Usually all blocks on a diagram are drawn the same size and all lines are drawn the same thickness. If special emphasis is desired, a larger block or a thicker line or a

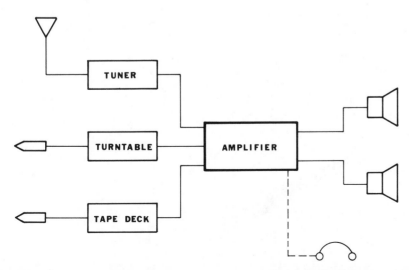

Figure 5-1 A block diagram of a stereo system.

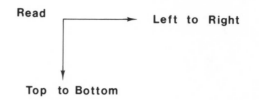

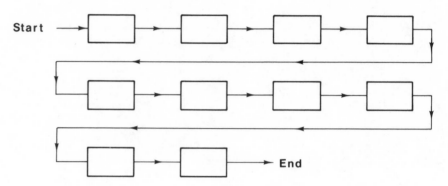

Figure 5-2 Block diagrams are read from left to right and top to bottom.

combination of both is drawn. Note how the amp in Figure 5-1 is emphasized by a thicker line.

Block diagrams are set up to be read from left to right and from top to bottom. If more than one line of blocks is required, the second line must be located under the first row and the line of flow must be returned from the end of the first row to the beginning of the second row as shown in Figure 5-2. If one of the blocks represents an element that is not normally part of the series, it can be drawn using hidden lines as was done with the earphones in Figure 5-1.

Block diagrams are usually drawn on paper printed with a nonreproducible gridded pattern (see Figure 5-3). This paper makes it much easier to layout equal-sized aligned blocks, and does not require the use of a scale. If gridded drawing paper is not available, a gridded board cover may be used.

Figure 5-3 Drawing paper with a non-reproducible grid.

Block diagrams are used not only to express graphically the relationship between a series of elements but can also be used to analyze a series of elements. Consider the following problem: You have a coin, one side of which is a head and the other side a tail, and you're going to flip the coin four times. What are the chances of getting a head four consecutive times and what are the chances of getting two heads and two tails in any order? Unless you are familiar with statistics, this problem can be very difficult to reason out. However, if you draw a block diagram showing the possible sequence of events, the problem can be easily solved. Figure 5-4 illustrates the four coin flips. We can see from Figure 5-4 that there are 16 possible combinations of heads and tails in four flips. Only one out of the 16 consists only of heads (combination #1). Therefore the chances of flipping 4 consecutive heads in 4 flips is 1 in 16.

Combinations 4, 6, 7, 10, 11, and 13 will yield combinations of two heads and two tails in any order, so the chances of flipping two heads and two tails in any order are 6 out of 16.

Figures 5-5 and 5-6 are further examples of block diagrams.

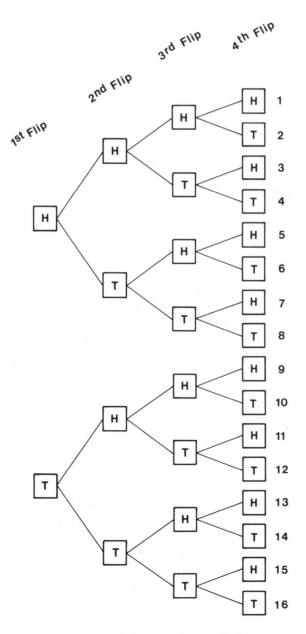

Figure 5-4 A block diagram which illustrates four flips of a coin.

FILTER CIRCUIT

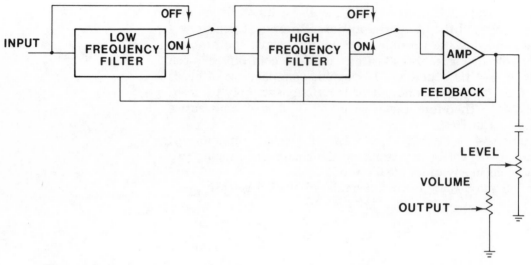

Figure 5-5 An example of a block diagram courtesy of General Electric Company.

Figure 5-6 An example of a block diagram courtesy of General Electric Company.

RECEIVER

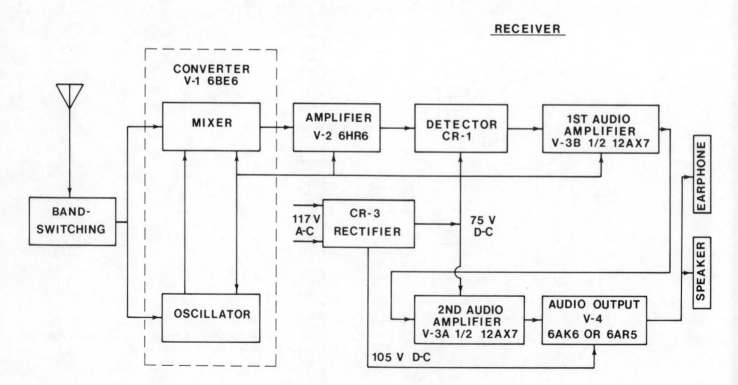

5-3 LOGIC DIAGRAMS

Logic diagrams are used to help design and analyze electronic circuits. They are particularly helpful in the analysis of circuits which are based on binary principles such as those found in computers and electronic calculators.

There are several different logic functions, each of which has its own logic symbol. The logic symbols are defined in Figure 5-7 and the dimensions for some of the symbols are presented in Figure 5-8. The sizes defined in Figure 5-8 are in concurrence with the Department of Defense standard MIL-STD-806 "Graphic Symbols for Logic Diagrams."

AND Function

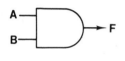

A	B	F
0	0	0
0	1	0
1	0	0
1	1	1

EXCLUSIVELY OR Function

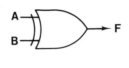

A	B	F
0	0	0
0	1	1
1	0	1
1	1	0

OR Function

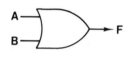

A	B	F
0	0	0
0	1	1
1	0	1
1	1	1

OTHER SYMBOLS

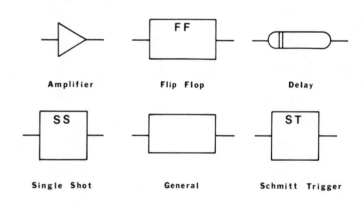

NAND Function (negative AND)

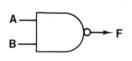

A	B	F
0	0	1
0	1	1
1	0	1
1	1	0

Figure 5-7 Graphic symbols for logic functions along with their truth tables.

NOR Function (negative OR)

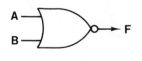

A	B	F
0	0	1
0	1	0
1	0	0
1	1	0

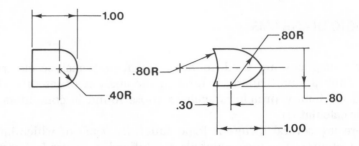

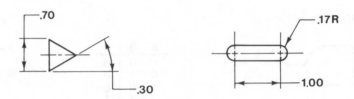

Figure 5-8 Recommended dimensions for various logic symbols.

These sizes are considered national standards, but may be varied so long as the basic shape remains the same.

Most draftsmen use templates as guides when drawing logic symbols. Figure 5-9 pictures one of the many different templates commercially available.

Logic diagrams are analyzed using *truth tables*. Figure 5-7 shows several basic logic functions and their truth tables. Truth tables are used to analyze not only individual logic functions but also those used in combination. For example, study the circuit shown in Figure 5-10. If the input is $A = 1$, $B = 0$, $C = 0$, then the AND function will generate 0 output. When this 0 output is combined with the $C = 0$ input into the OR function, the final output will be 0. If we change the inputs to $A = 0$, $B = 0$, $C = 1$, the AND function will produce a 0 output. This 0 coupled with the $C = 1$ input will, when acted upon by the OR function, generate a final output of 1. The truth table shown in Figure 5-10 analyzes all the possible combinations for the circuit.

Figure 5-9 A template which includes logic symbol cutouts.

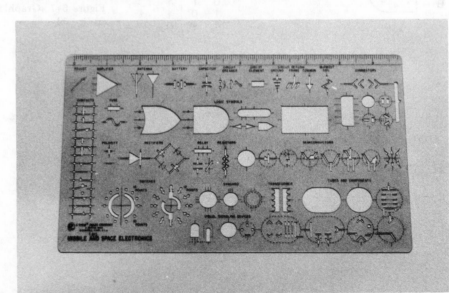

Sometimes it's difficult to remember all the possible combinations that a number of inputs could generate. Figure 5-11 has been included to help define all the possible combinations for up to 5 given inputs.

Figures 5-12 and 5-13 are examples of logic diagrams along with their truth tables. Study each one and verify the meaning of each set of input combinations.

Figure 5-14 compares some logic diagrams with their equivalent schematic representations.

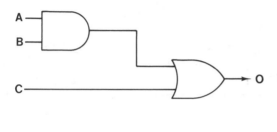

A	B	C	O
0	0	0	0
0	0	1	1
0	1	0	0
0	1	1	1
1	0	0	0
1	0	1	1
1	1	0	1
1	1	1	1

Figure 5-10 An example of a logic diagram along with its truth table.

E	D	C	B	A	
0	0	0	0	0	
0	0	0	0	1	
0	0	0	1	0	2 Inputs
0	0	0	1	1	
0	0	1	0	0	
0	0	1	0	1	
0	0	1	1	0	3 Inputs
0	0	1	1	1	
0	1	0	0	0	
0	1	0	0	1	
0	1	0	1	0	
0	1	0	1	1	
0	1	1	0	0	
0	1	1	0	1	4 Inputs
0	1	1	1	0	
0	1	1	1	1	
1	0	0	0	0	
1	0	0	0	1	
1	0	0	1	0	
1	0	0	1	1	
1	0	1	0	0	
1	0	1	0	1	
1	0	1	1	0	
1	0	1	1	1	
1	1	0	0	0	
1	1	0	0	1	
1	1	0	1	0	
1	1	0	1	1	
1	1	1	0	0	
1	1	1	0	1	
1	1	1	1	0	5 Inputs
1	1	1	1	1	

Figure 5-11 Possible combinations for up to five inputs. Note that for 2 inputs there are 4 possible combinations. For 3 inputs 8 combinations, 4 inputs 16 combinations, 5 inputs 32 combinations, etc.

A	B	C	O
0	0	0	0
0	0	1	0
0	1	0	0
0	1	1	0
1	0	0	1
1	0	1	1
1	1	0	1
1	1	1	0

Figure 5-12 An example of a logic diagram along with its truth table.

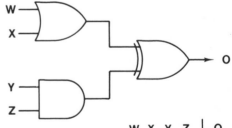

W	X	Y	Z	O
0	0	0	0	0
0	0	0	1	0
0	0	1	0	0
0	0	1	1	1
0	1	0	0	1
0	1	0	1	1
0	1	1	0	1
0	1	1	1	0
1	0	0	0	1
1	0	0	1	1
1	0	1	0	1
1	0	1	1	0
1	1	0	0	1
1	1	0	1	1
1	1	1	0	1
1	1	1	1	0

Figure 5-13 An example of a logic diagram along with its truth table.

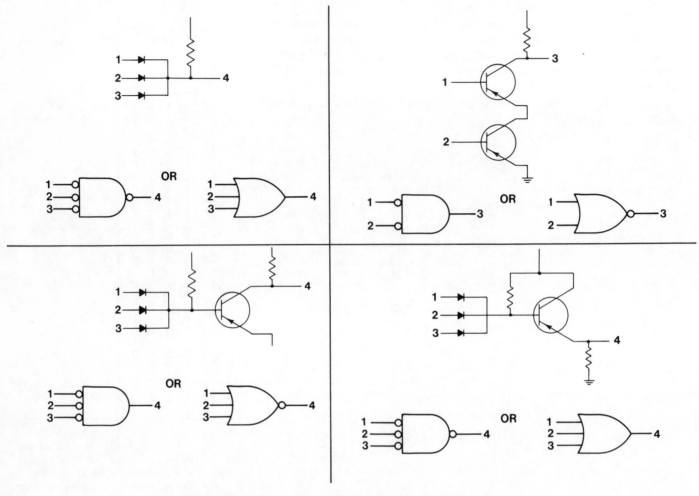

Figure 5-14 Some logic diagrams with their equivalent schematic diagrams.

PROBLEMS

5-1 Redraw the block diagram shown in Figure 5-1 and add two more speakers.

5-2 Prepare a block diagram which displays the following series of events.

	Get up
	Go to school
	Go to class
	Go to lunch
Go to class	Play cards
Go home	Play more cards
Study	Go home
	Eat supper

5-3 Redraw, using instruments, the block diagram shown in Figure 5-15.

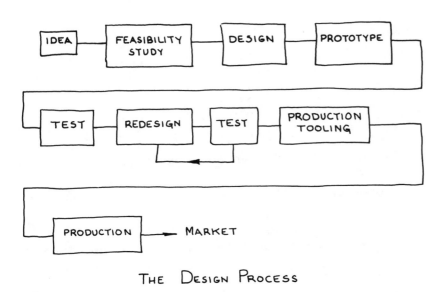

THE DESIGN PROCESS

Figure 5-15

5-4 Redraw the logic diagrams in Figure 5-16 and complete the truth tables.

Figure 5-16

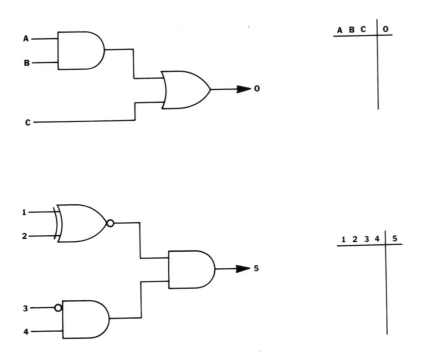

5-5   Redraw the following truth table and add a logic diagram whose functions will generate the truth table.

A	B	C	Output
0	0	0	0
0	0	1	0
0	1	0	0
0	1	1	0
1	0	0	0
1	0	1	0
1	1	0	0
1	1	1	1

5-6   Redraw the logic diagram shown in Figure 5-17 and add the truth table.

5-7   Redraw the logic diagrams shown in Figure 5-18 and add the truth table.

5-8   Redraw the logic diagram shown in Figure 5-19 and complete a truth table for any 6 of the 64 possible input combinations.

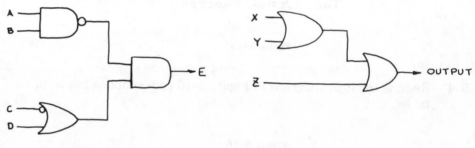

Figure 5-17

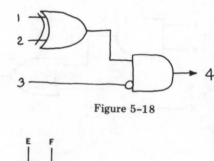

Figure 5-18

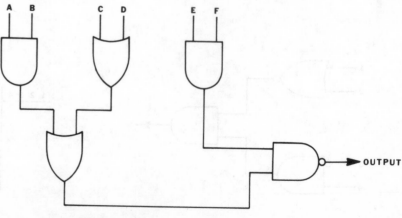

Figure 5-19

PC BOARD
DRAWINGS

6-1 INTRODUCTION

This chapter deals with how to prepare PC Board drawings, drilling drawings, and PC assembly drawings. In addition, it will include an explanation of tape and inking techniques as they apply to PC board drawings.

6-2 HOW TO PREPARE A PC BOARD DRAWING

Figure 6-1 is an example of a PC board drawing. The procedure used to create PC board drawings is as follows and is illustrated in Figure 6-2.

Figure 6-1   A PC board drawing.

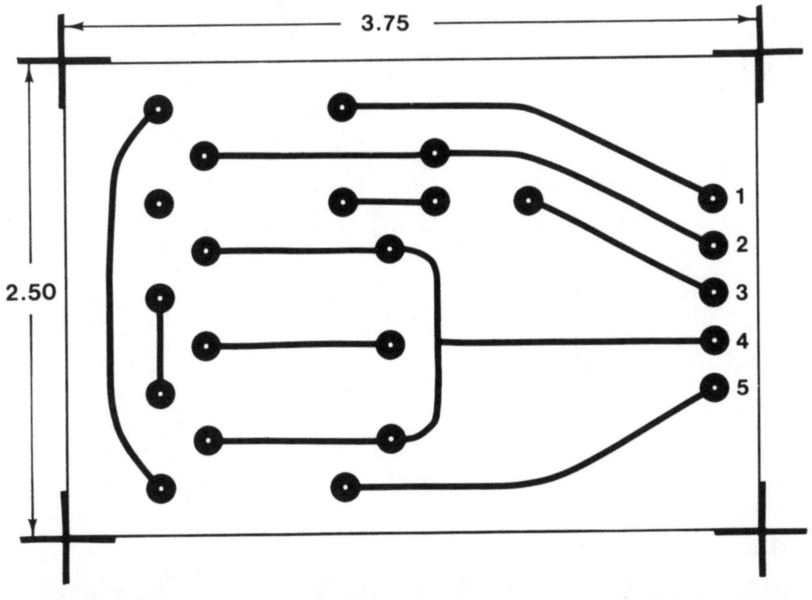

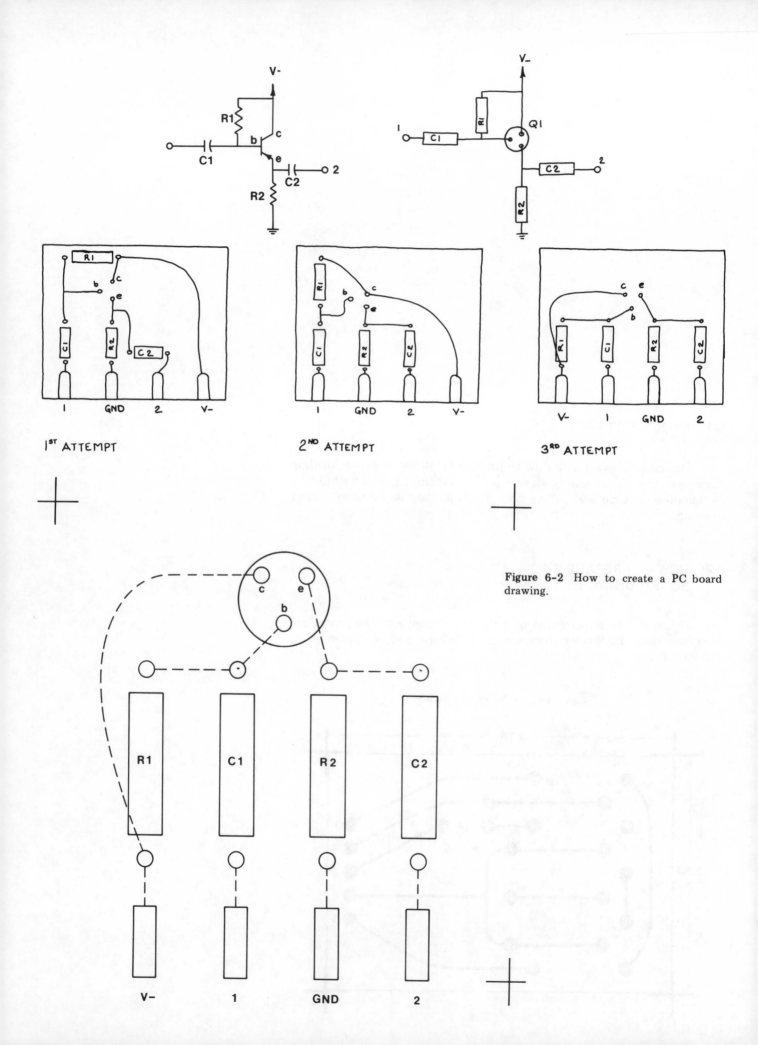

1ST ATTEMPT

2ND ATTEMPT

3RD ATTEMPT

Figure 6-2 How to create a PC board drawing.

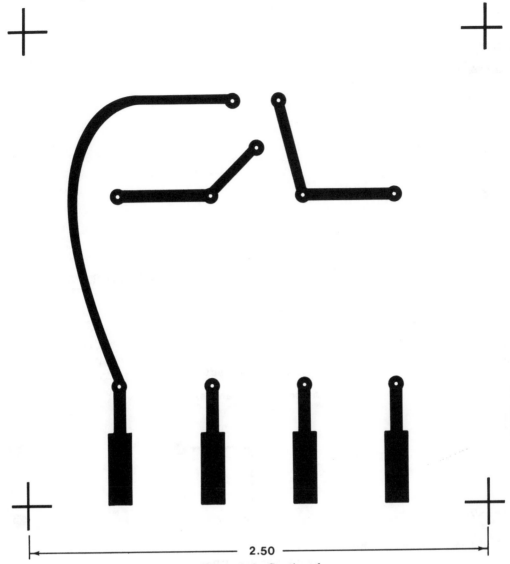

2.50

Figure 6-2 Continued.

1. Study the schematic diagram and identify all components.

In our example we have two resistors, two capacitors, a transistor, and four inputs: 1, 2, V-, and a ground.

2. Redraw the schematic diagram using pictorial symbols.

This step is optional but is helpful for beginners. It helps translate the symbols into their actual shapes while maintaining the correct conductor paths.

The component sizes are available from manufacturers catalogs. For our example we can assume that both the resistors and capacitors are 0.75 long and 0.19 dia. and that the transistor is 0.50 dia.

3. Using trial and error, determine the optimum component arrangement and conductor path pattern. This step is difficult to complete because there is no exact "optimum" arrangement. There are always several very good alternatives.

Figures 6-3 and 6-4 illustrate several guidelines which will help in determining component arrangement and conductor path pattern.

In our example, three attempts were made before an acceptable arrangement was found. The first attempt located the four inputs along

General conductor path sizes – will vary with voltage

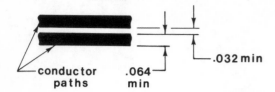

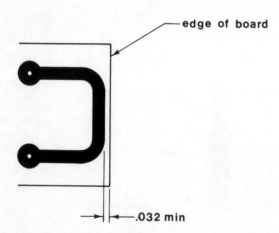

edge of board

.032 min

Figure 6-3 Guidelines for drawing conductor paths.

CONDUCTOR PATHS

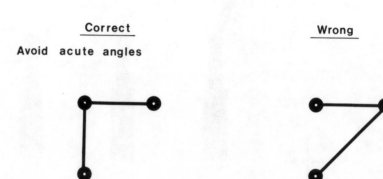

Correct   Wrong

Avoid acute angles

Avoid sharp turns

Keep paths as short as possible

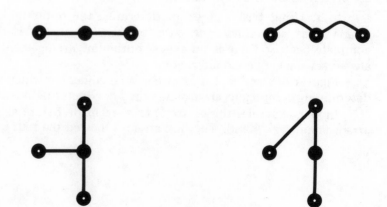

Best   arrangement

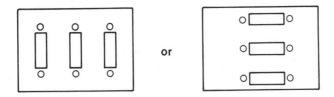

An acceptable arrangement

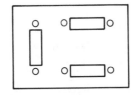

Avoid !!

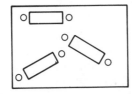

Figure 6-4   How to arrange components on a PC board.

one edge of the board but located R1 and C2 horizontally and C1 and R2 vertically. The second attempt aligns the components all in the same direction but has excessively long conductor paths. The third attempt shortens the conductor paths by rotating the transistor 90° counterclockwise, relocating R1, and by locating the V-terminal pad.

4. Using the final sketch of step 3, prepare the PC board drawing.

PC board drawings are usually drawn "up," that is, at an increased scale. Scales of 4 = 1 or 5 = 1 are most often used.

For our example, a scale of 4 = 1 was chosen. This means that the components are drawn four times larger than they actually are. The resistors and capacitors are drawn 3″ long and ¾″ wide. The transistor is drawn as a 2″ diameter circle. As you layout the components on the PC board, try to coordinate the locations so that they match the grid on which you are drawing. In our example, drawn at a scale of 4 = 1, all components were centered on the inch lines of the grid pattern. This centering helps to simplify the manufacturing process.

Components are drawn using solid (visible) lines and conductor paths are drawn using hidden lines. In other words, the board is drawn with the component side up. The conductor paths are on the bottom of the board.

The outside corners of the board are defined by the four large plus signs. The actual edges may be drawn but are not necessary. Only the corners need be defined.

The distance between the holes used to mount the resistors and capacitors is determined by the formula (see Figure 6-5):

$$MD = R + 8t \qquad\qquad (F1)$$

where

MD = Mounting distance
R = Component length
t = Wire thickness

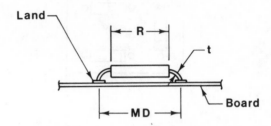

Figure 6-5 Definition of terms used in formula F1.

However, as we located the components to align with the background grid pattern, so we also try to align mounting holes. Formula (F1) serves as an approximation which we can then round off. If, for example, we calculated a mounting distance to be 0.72, on an eight-to-the-inch pattern it would be drawn 0.75 and on a ten-to-the-inch pattern it would be drawn 0.70.

For our illustrated example, MD calculates to be 0.996 (assuming $t = 0.032$). At a scale of 4 = 1 this is 3.98. Align this to the grid pattern and we find the final mounting distance to be 4 inches or on the PC board.

5. Prepare the tape drawing.

Figure 6-2 illustrates the drawing. The conductor paths are drawn 0.25 wide and the mounting pads 0.38 dia.

The only dimension that need appear on a taped drawing is the width. This one dimension is in turn used to calibrate the final board size. The dimension should define the finished size of the board, not the "up" size.

6. If required, prepare a drilling drawing for the board.

Figure 6-6 illustrates a drilling drawing, which defines where and what diameter holes are to be drilled into the board. The illustrated example dimensions the hole pattern using the coordinate dimensioning system which is explained in Section 7-5.

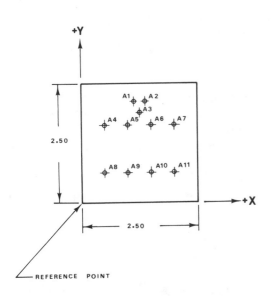

HOLE	X	Y	DIA
A1	1.12	2.25	.032
A2	1.38	2.25	↑
A3	1.25	1.88	
A4	0.50	1.62	
A5	1.00	↑	
A6	1.50	↓	
A7	2.00	1.62	
A8	0.50	0.62	
A9	1.00	↑	
A10	1.50	↓	↓
A11	2.50	0.62	.032

Figure 6-6 A drilling drawing for the PC board designed in Fig. 6-2.

6-3 TAPING TECHNIQUES

Many PC drawings are created using tape. Tape is used because it is very durable, can easily be removed and relocated without damaging the drawing medium, and produces a very black, dense drawing which is ideal for the photographic process used to manufacture PC boards.

Figure 6-7 shows several different types of pre-cut tape along with a roll of $\frac{1}{16}$-inch-wide tape. The circle and rectangular shapes are used for terminals and are called pads. The roll of tape is used for conductor

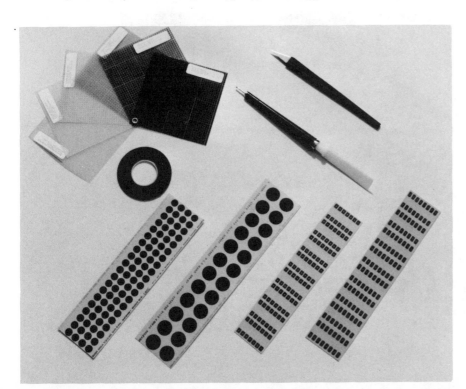

Figure 6-7 Pre-cut tape for help in creating PC board drawings. Courtesy of W.H. Brady Co.

paths. The brandishing pen is used to rub the tape into place and the exacto knife is used to cut the tape. The exacto knife can also be used to scrape the tape off if it has been put on the drawing incorrectly.

Also shown in Figure 6-7 is a set of gridded mylar samples. Mylar is a very durable plastic film which is an excellent drawing medium for use with tape.

Tape is applied to the drawing medium by first locating and then rubbing. A blunt tool, such as a burnishing pen, the unpointed end of a ballpoint pen, or the unpointed end of a leadholder, may be used to rub the tape firmly into place. Figure 6-8 shows precut drafting aids being applied.

Pre-cut tape drafting aids are available in a wide variety of shapes and sizes. They enable a draftsman to work much more quickly and

Figure 6-8 How to apply pre-cut tape to create a PC drawing. Courtesy of W.H. Brady Co.

more accurately than is possible with ink or pencil techniques. They are, however, expensive and should be used carefully.

6-4   INKING TECHNIQUES

Inked PC drawings are sometimes prepared because they are extremely durable and are well suited to photographic manufacturing techniques. Ink drawings are prepared using technical drawing pens such as the one shown in Figure 6-9.

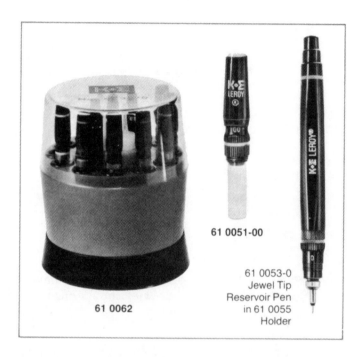

61 0051-00

61 0053-0
Jewel Tip
Reservoir Pen
in 61 0055
Holder

61 0062

Figure 6-9 Technical drawing pens. Courtesy of B.L. Makepeace Co., Boston, Mass.

Technical pens come in various widths labeled from 000 to 5 with 000 being the narrowest line. A pen is only able to produce one size line, so a different line width will require a different pen. Draftsmen who do inking work usually carry several different sized pens.

Two rules should be followed when preparing inked drawings: hold the pen perfectly vertical, allowing the ink to flow out of the pen and always allow an air gap between the guiding instrument and the drawing medium. Unlike a ballpoint pen, or a fountain pen, a technical pen requires very little pressure to operate. Once a line is started, the ink will continue to flow until the pen is lifted from the drawing. If a pen is simply stopped at the end of a line, a keyhole effect will result.

An air gap, see Figure 6-10, is required to keep the ink from flowing under guiding instruments. An air gap is created by lifting the instrument off the drawing surface by sliding a triangle or template under the instrument to be used for the drawing. It may also be done by sticking pressure pads (circular pieces of plastic with glue on one side) or by taping pennies to the drawing instrument.

Figure 2-7 pictures a Leroy inking setup. Leroy is a tracing technique which uses a bar with symbols engraved and a scriber. The scriber

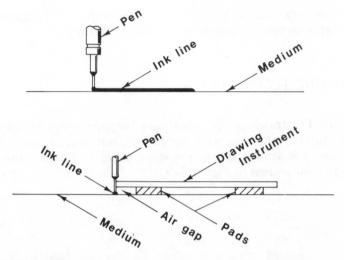

Figure 6-10 Drawing instruments must be off the drawing surface when inking.

has two points and an ink well. One point rides in a groove at the bottom of the bar and references the scriber. The second point is used to trace the engraved symbols, and the ink well in turn follows the second point copying the traced symbol onto the drawing. A technical pen can be used in place of the inkwell.

PROBLEMS

6-1 Prepare a PC board drawing of the schematic diagram shown in Figure 3-16. The final drawing should include:

  a. A pictorial schematic of the circuit.
  b. Freehand sketches of the various attempts to achieve an acceptable component arrangement.
  c. A final layout drawn at a scale of 4 = 1 (a 2 = 1 scale may be employed in order to permit a smaller paper size to be used).
  d. A final PC drawing at a scale of 4 = 1. If tape or ink is not available, use pencil shading.
  e. Prepare a drilling drawing at a scale of 1 = 1.

Assume all inputs, outputs, grounds, power sources, are 0.20 × 0.80 tabs and are all located on one edge of the board. Assume all resistors and capacitors are 0.50 long, 0.20 dia., and have 20 gage mounting wires. Keep the board size as small as possible. Assume the transistor is 0.50 in diameter.

6-2 Repeat Problem 6-1 for Figure 3-17. Assume the speaker is connected to the board via the end tabs.

6-3 Repeat Problem 6-1 for Figure 3-18.

6-4 Repeat Problem 6-1 for Figure 3-3.

CHASSIS
DRAWINGS

7-1 INTRODUCTION

This chapter deals with the design and drawing of chassis. It includes an explanation of the different drawing techniques used to draw chassis, an explanation of how chassis flat patterns are developed, and a review of some of the fasteners used in chassis design.

7-2 TYPES OF CHASSIS

There are many different materials and shapes which are used to create a wide variety of chassis designs. General categories of chassis designs are usually identified by the basic shape. For example: box, U, T, and I chassis are designs which are shaped like a box, U, T, or I respectively. Figure 7-1 shows a U type chassis with flanges.

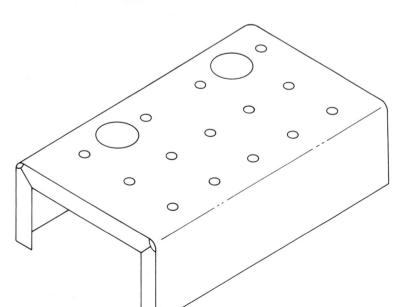

Figure 7-1   A U type chassis with flanges.

Chassis may be purchased from commercial manufacturers or they may be fabricated in a local shop.

7-3   FLAT PATTERN DESIGN

Chassis are fabricated by first cutting a flat pattern from the desired material and then bending the flat pattern into the final chassis shape. As metal is bent, it stretches, which means that the length of the flat pattern must be less than the finished overall length of the chassis.

The correct size of a flat pattern is determined by considering the overall size requirements and the amount of material needed for "bend allowance." Bend allowance is the amount of material needed to form bends and includes a consideration of how the material will stretch during the bending. Bend allowances and the lengths of straight sections are added together to determine the diminished length of the flat pattern.

Before proceeding with a discussion of how to determine flat pattern lengths, we must first understand three terms: inside bend radius, outside bend radius, and neutral axis. Figure 7-2 defines the terms. The neutral axis is a theoretical line exactly half-way between the inside and outside bend radius. Material between the neutral axis and the inside bend radius is under compression during bending and material between the neutral axis and the outside bend radius is under tension (being stretched).

The difference between the inside and outside bend radii is equal to the material thickness.

Inside bend radius + Material thickness = Outside bend radius (F-1)

Drawings usually only define inside bend radius and material thickness, so if the value of the outside bend radius is desired, it must be calculated using formula (F1). In Figure 7-3 the inside bend radius is $^3/_{16}$ and the material thickness is $^1/_8$. Applying these data to formula (F1) we get

$$^3/_{16} + ^1/_8 = \text{Outside bend radius}$$

$$^3/_{16} + ^2/_{16} = \text{Outside bend radius}$$

$$^5/_{16} = \text{Outside bend radius}$$

Figure 7-2 Illustration of inside bend radius, neutral axis, outside bend radius.

Figure 7-3

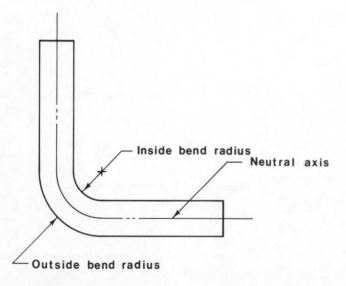

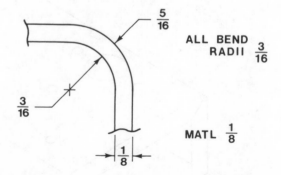

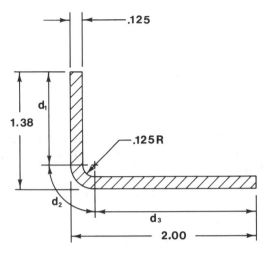

Figure 7-4

The length of a flat pattern is determined by adding the length of the straight sections of the final chassis shape to the lengths of the curved sections which have been modified to account for bend allowance. If we apply this concept to the contour shown in Figure 7-4, we see that

$$\text{Flat pattern length} = d_1 + d_2 + d_3$$

The distance d_1 is found by taking the overall length and subtracting the inside bend radius and material thickness.

$$d_1 = 1.38 - 0.125 - 0.125$$
$$d_1 = 1.13$$

Likewise d_3

$$d_3 = 2.00 - 0.125 - 0.125$$
$$d_3 = 1.75$$

The distance d_2 must be calculated considering bend allowance. This is done by the formula:

$$B = \frac{A}{360}\, 2\pi(1R + Kt)* \qquad\qquad (F2)$$

where

B = Bend allowance

A = Bend angle

IR = Inside bend radius

t = Material thickness

K = Constant

*From ASTME Die Design Handbook.

For the problem presented in Fig 7–4,

$$A = 90°$$

$$IR = 0.125$$

$$t = 0.125$$

$$K = 0.33$$

The value of K is equal to 0.33 when the inside bend radius is less than $2t$ (two times the material thickness) and is equal to 0.50 when inside bend radius is greater than $2t$. In our problem,

$$2t = 2(0.125) = 0.25$$

which means

$$K = 0.33$$

as the inside bend radius is 0.125 which is less than 0.25. Substituting the values in formula (F2)

$$B = \frac{90}{360}(2\pi)[0.125 + (0.33)(0.125)]$$
$$= 1.57(0.125 + 0.041)$$
$$B = 0.26$$
$$\therefore d_2 = 0.26$$

We can now calculate the flat pattern length by adding d_1, d_2, and d_3:

$$\text{Flat pattern length} = d_1 + d_2 + d_3$$
$$= 1.13 + 0.26 + 1.75$$
$$= 3.14$$

Figure 7–5 is another example of how to calculate the flat pattern length for a given chassis shape.

7-4   HOW TO DRAW CHASSIS

Chassis are sometimes difficult to draw because they contain few square corners and are made from thin materials, which makes drawing hidden lines confusing. These difficulties may be overcome by applying the following drawing techniques.

Round corners are easier to draw if we first draw the object as shown in Figure 7–6. Then darkening in the object lines, start by first darkening in the round corners. This will make it easier to assure an even, continuous line both into and out of the corner. Also, remember when drawing round corners that the labels on a circle template are diameters, whereas bending values are defined in radii. Don't forget to multiply radius values by 2 before using a circle template.

$$2(\text{radius}) = \text{diameter}$$

Hidden lines are drawn in sheet metal parts as shown in Figure 7–7. In every case, the centerline for a hole is always included. If the

STRAIGHT SECTIONS

VERTICAL (2 REQD) = .75 - .062 - .19

= .498

×2 ⟶ .996

HORIZONTAL = 2.25 - .062 - .062 - .19 - .19

= 1.745 ⟶ 1.745

CURVED (2 REQD) = $\frac{A}{360} 2\pi (1R + Kt)$

= $\frac{90}{360} 2\pi [.19 + (.50)(.062)]$

K = .50 BECAUSE

.19 IS GREATER THAN

2(.062) = .124

= 1.57 (.19 + .031)

= .347

×2 ⟶ .693

3.434

Bend Radius = .19
Thickness = .062

.75

2.25

Figure 7-5 An example of a flat pattern length calculation.

FLAT PATTERN LENGTH

= 3.434

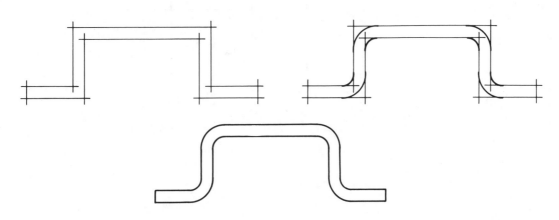

Figure 7-6 How to draw shapes made from thin materials.

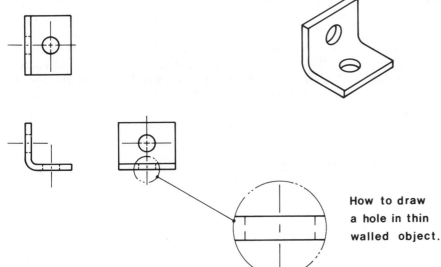

How to draw a hole in thin walled object.

Figure 7-7 Holes in thin materials.

material is so thin that the convention pictured in Figure 7-7 is not practical, an enlarged detail should be considered.

7-5   HOW TO DIMENSION CHASSIS DRAWINGS

Chassis designs are characterized by many small holes in a small area. Conventional dimensioning techniques would tend to create drawings which would be a maze of numbers, extension, dimension, and leader lines. To help produce neat, clear, and easy-to-read chassis drawings, two different dimensioning systems are commonly used: the baseline and the coordinate systems.

The baseline system refers all dimensions to common baselines. These lines are sometimes called datum lines or reference lines. The baseline system is illustrated in Figure 7-8. The baseline system has an advantage of eliminating cummulative tolerance errors. Each dimension

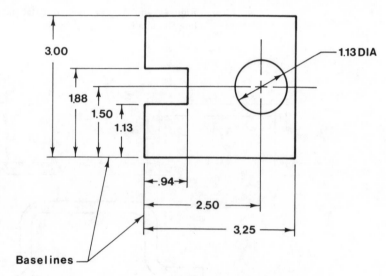

Figure 7-8   An example of baseline system dimensioning.

is taken independently and an error in one dimension will not carry over to other dimensions. The disadvantages of the baseline system are that it requires a large amount of area to complete, usually at least twice the area of the object, and that once the dimensions are in place on the drawing, it is very difficult to make changes or additions.

Most chassis have rounded edges which do not make good dimensioning reference lines because they are difficult for the craftsman to consistently align. Most draftsmen use a centerline or some other line within the object as a reference line. Note the location of the baselines in Figure 7-9.

To use the baseline system (see Figure 7-10):

1. Prepare a scale drawing of the chassis surface locating all holes. Be sure to include centerlines for every hole.

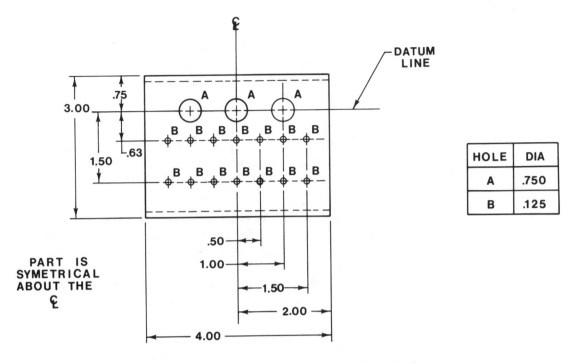

Figure 7-9 An example of baseline system dimensioning.

HOLE	DIA
A	.750
B	.125

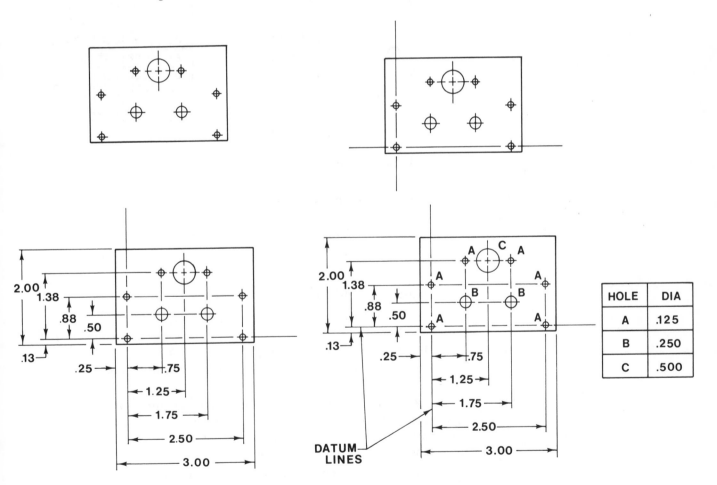

Figure 7-10 How to dimension a chassis surface using the baseline system.

HOLE	DIA
A	.125
B	.250
C	.500

2. Define the baselines. Each surface will require two baselines.

3. Locate all holes and edges from the baselines. Space the dimension lines ¼″ apart and place dimensions in consecutive order according to length starting with the shortest dimension closest to the baseline.

4. Define the hole sizes by assigning each different hole size a letter and then defining each letter in a chart as shown in Figure 7-10. Place the letters on the drawing to the right and above the appropriate hole if possible. Keep the letters close to the holes they define.

The coordinate system is based on a 90° *x-y* coordinate system and is particularly well suited for use when programming numerically controlled machines. The coordinate system is easier to draw than the baseline system and is easier to change or correct but does require the reader to look up each dimension. Figure 7-11 illustrates a chassis surface dimensioned using the coordinate system.

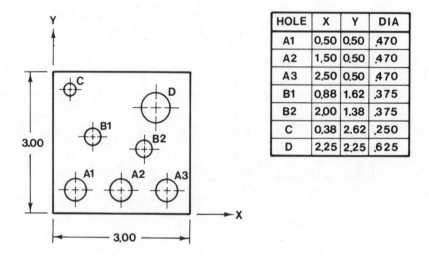

HOLE	X	Y	DIA
A1	0.50	0.50	.470
A2	1.50	0.50	.470
A3	2.50	0.50	.470
B1	0.88	1.62	.375
B2	2.00	1.38	.375
C	0.38	2.62	.250
D	2.25	2.25	.625

Figure 7-11 An example of coordinate system dimensioning.

To use the coordinate system (see Figure 7-12):

1. Prepare a scale drawing of the chassis surface locating all holes. Be sure to include centerlines for every hole.

2. Define the reference point and then clearly label the positive *x* and *y* axis. Add the overall dimensions.

3. Label each hole using a letter and a number. Assign holes of equal diameter the same letter, but different consecutive numbers. In the example problem, the holes labeled *A* are all 4 mm in diameter; holes labeled *B* are 10 mm in diameter, etc.

4. Prepare a chart which includes a listing, using the labels assigned in step 3, of the distance each hole is from the *x* and *y* axis and the hole's diameter.

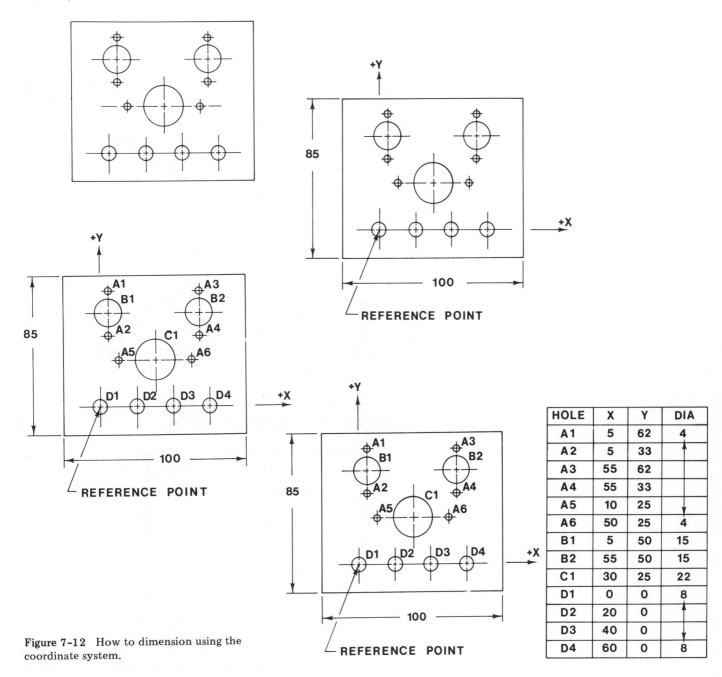

Figure 7-12 How to dimension using the coordinate system.

HOLE	X	Y	DIA
A1	5	62	4
A2	5	33	
A3	55	62	
A4	55	33	
A5	10	25	
A6	50	25	4
B1	5	50	15
B2	55	50	15
C1	30	25	22
D1	0	0	8
D2	20	0	
D3	40	0	
D4	60	0	8

7-6 CHASSIS FASTENERS

There are several different types of fasteners which are frequently used to manufacture chassis including rivets, sheet metal screws, machine screws. Welded and soldered joins are also used. Thread representations were covered earlier in Section 1–10 and these representations are also valid for threaded chassis fasteners.

Rivets are shown on a drawing by using the representations shown in Figure 7–13. There are two types of representations: detailed and

Detailed   **Schematic**

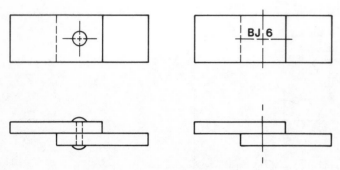

Figure 7-13 How to draw rivets.

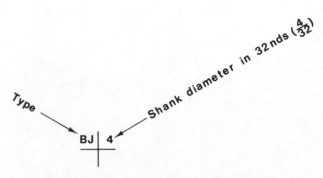

Figure 7-14 Definition of rivet notations.

schematic. Rivets can be identified by using the coding system of the National Aircraft Standards (NAS) which is illustrated in Figure 7-14.

A long row of rivets, provided that the rivets are all exactly the same kind, may be called out by calling out only the first and last rivet in the row. Figure 7-15 illustrates this kind of rivet callout.

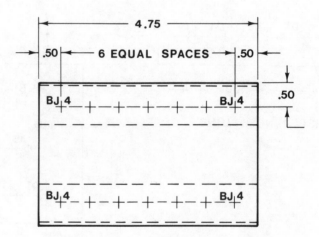

Figure 7-15 How to call out a row of rivets.

Spot welds are shown on a drawing using the symbols presented in Figure 7-16. They may be represented by either of the symbols shown. The dimensions for the symbol sizes are only guidelines and need not be strictly followed.

Spot welds are sometimes more advantageous in a design because they do not require any space, whereas a rivet requires space for its head and butt end. Rivets, on the other hand, may be drilled out and replaced whereas a spot weld must be rewelded. Rewelding usually requires almost complete disassembly of a chassis.

The desired diameter of a spot weld may be indicated by lettering in the diameter size to the left of the circle on the weld symbol. The distance between spot welds, as measured from center to center and called the "pitch" of the spot welds, may be indicated by lettering in the distance value to the right of the circle on the symbol (see Figure 7-16).

SPOTWELDS

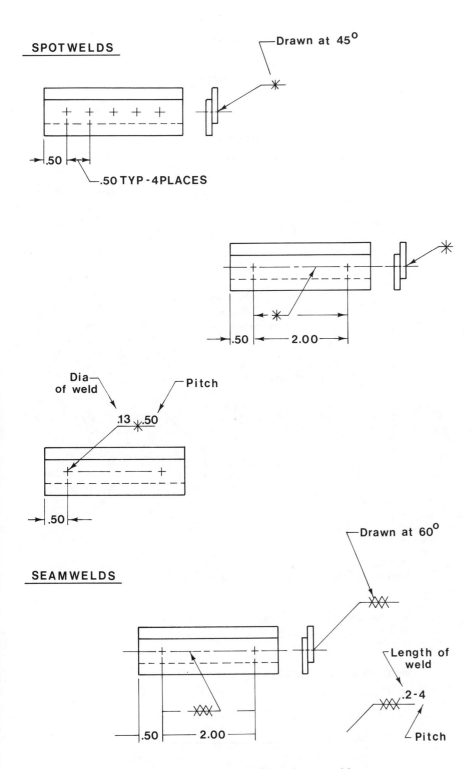

SEAMWELDS

Figure 7-16 Spotwelds and seamwelds.

PROBLEMS

7-1 Draw and dimension the flat pattern for the shapes defined in Figure 7-17.

a.

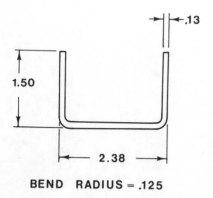

b.

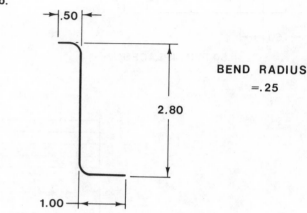

c.

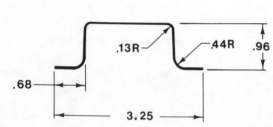

d. (Millimeters)

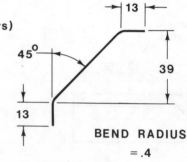

Figure 7-17

7-2 Dimension the chassis surface shown in Figure 7-18 using:

a. the coordinate system
b. the baseline system

Each block on the grid pattern equals 0.125.

Figure 7-18

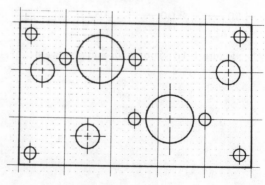

116

7-3   Dimension the chassis surface shown in Figure 7-19 using:

    a. the coordinate system
    b. the baseline system

Each block on the grid pattern equals 0.20.

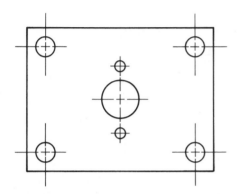

Figure 7-19

ALL DIMENSIONS TO THE NEAREST .10

Figure 7-20

7-4   Dimension the chassis surface shown in Figure 7-20 using:

    a. the coordinate system
    b. the baseline system

Each block on the grid pattern equals 0.20.

7-5   Draw and dimension a flat pattern for the chassis shape shown in
Figure 7-21. The holes are to be drilled before the material is
bent. Each block on the grid pattern equals 0.20.

Figure 7-21

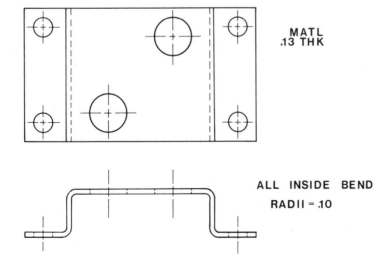

MATL
.13 THK

ALL INSIDE BEND
RADII = .10

7-6 Redraw the figure shown in Figure 7-22 and identify the fasteners as

a. BJ6 rivets
b. $^3/_{16}$ dia. spot welds
c. remove all fasteners and use a seam weld.

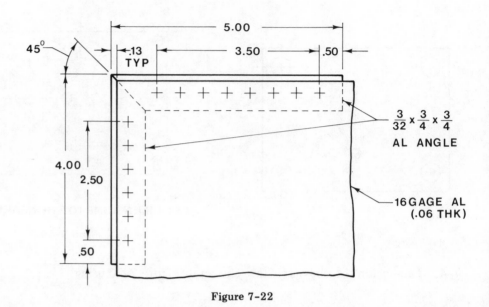

Figure 7-22

PICTORIAL DRAWINGS

8-1 INTRODUCTION

Electrical and electronic draftsmen are often asked to prepare a pictorial drawing of the project they are working on. These pictorial drawings may be used as part of a sales presentation to help customers better understand what they are purchasing, or they may be used in assembly or maintenance instructions to make it easier for the crafts-persons and mechanics to understand the instructions.

There are many different types of pictorial drawings, but the isometric and oblique types are most often used. Figure 8-1 illustrates the two types of drawings.

Isometric drawings are the more visually correct of the two, but are the more difficult to draw. Oblique drawings are easier to draw, but present a minimum pictorial representation.

8-2 ISOMETRIC DRAWINGS

Isometric drawings are pictorial drawings which are based on an isometric axis: two 30° lines and one vertical line (see Figure 8-2). Customarily the left isometric plane is used to show the front view of the object, the right plane the right side view, and the top plane the top view. Figure 8-3 illustrates this convention.

To prepare an isometric drawing, use the following procedure, which is illustrated in Figure 8-4:

1. Layout an isometric axis.

2. Draw the length, depth, and width of the object as shown.

3. Add, using very light construction lines, the preliminary detail of the object.

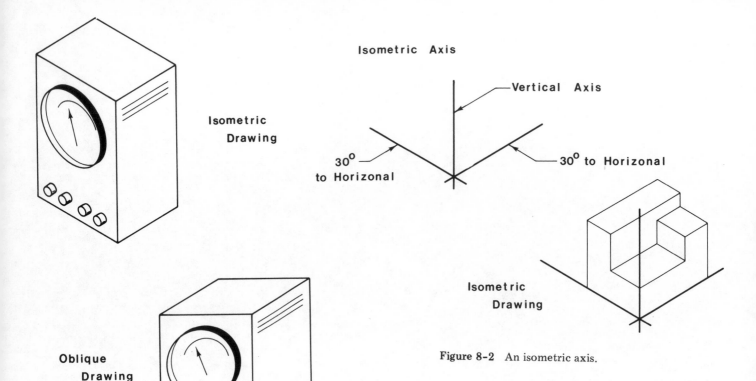

Isometric Drawing

Isometric Axis

Vertical Axis

30° to Horizonal

30° to Horizonal

Isometric Drawing

Figure 8-2 An isometric axis.

Oblique Drawing

Figure 8-1 An isometric and oblique drawing of a meter.

Figure 8-3 The left isometric plane is used to show the front view, the right plane the right side view, and the top plane the top view.

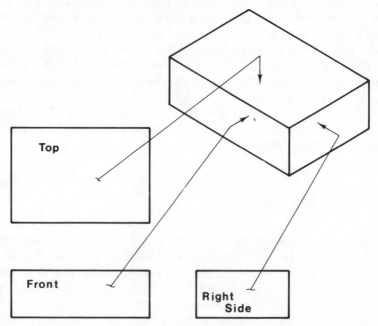

4. Add the final detail to the drawing.

5. Erase any excess lines and draw, using heavy black lines, the final shape of the object. The final drawing may be traced if the layout created in step 4 can not be erased cleanly.

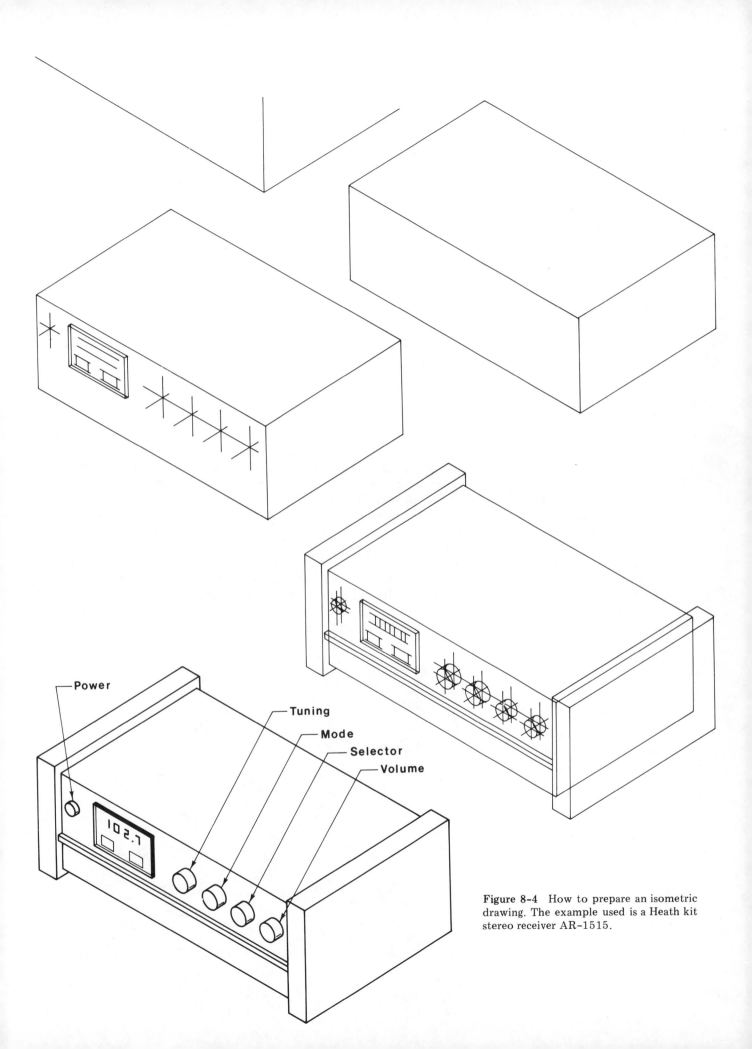

Power

Tuning

Mode

Selector

Volume

102.7

Figure 8-4 How to prepare an isometric drawing. The example used is a Heath kit stereo receiver AR-1515.

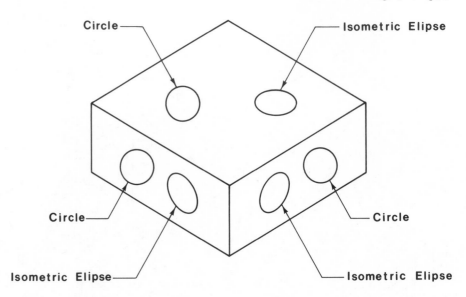

Figure 8-5 Isometric drawing requires ellipses to represent holes. Circles will appear distorted.

One shape that usually causes beginners difficulty when preparing isometric drawings is a circle. This is because circles are not drawn as circles but as ellipses. Figure 8-5 illustrates the visual distortion that circles create when drawn on isometric drawings and shows the visually correct picture ellipses create.

Correctly shaped ellipses can easily be drawn on isometric drawings by using an isometric ellipse template. Figure 8-6 pictures an isometric ellipse template and demonstrates how the template is used. For ellipses in the left-hand plane, rotate the template as shown so that the edge labeled LEFT-HAND PLANE rests on the T-square. For ellipses in the right plane use the edge labeled RIGHT-HAND PLANE as the base, and for ellipses in the top plane use the edge labeled HORIZONTAL PLANE as a base.

Figure 8-6 How to use an isometric ellipse template.

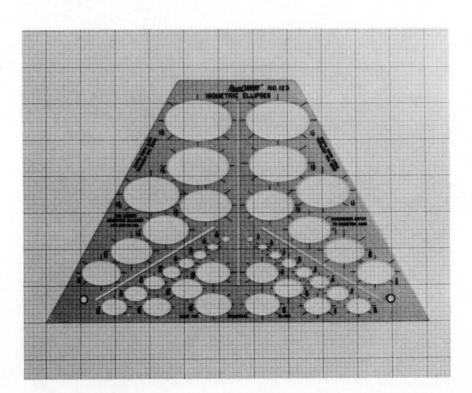

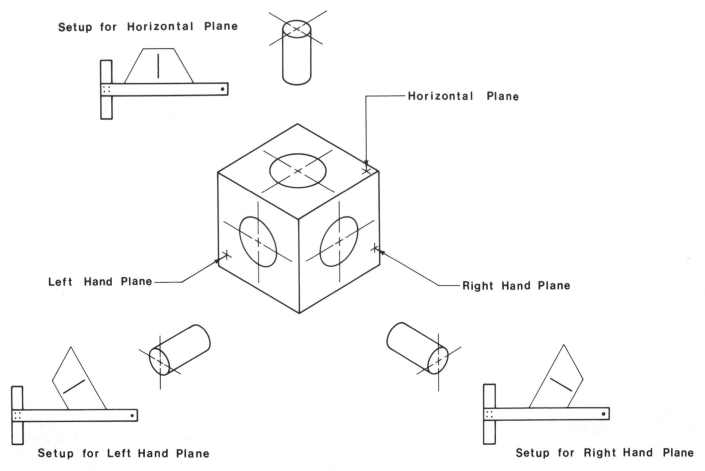

Setup for Horizontal Plane

Horizontal Plane

Left Hand Plane

Right Hand Plane

Setup for Left Hand Plane

Setup for Right Hand Plane

Figure 8-6 Continued.

Several other templates are used for preparing isometric drawings. An isometric protractor is helpful when marking off dial indications or meter scales as shown in Figure 8-7. A small ellipse template, pictured in Figure 8-8, is used to help draw very small ellipses which are not included on the isometric ellipse template previously discussed.

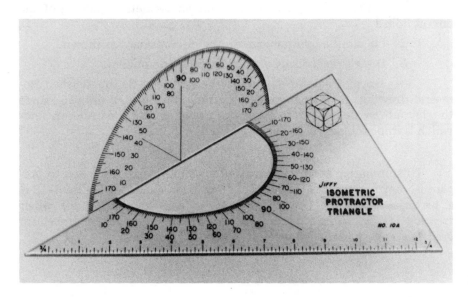

Figure 8-7 An isometric protractor.

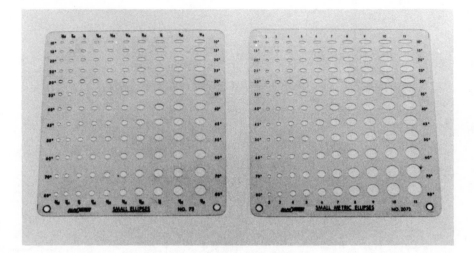

Figure 8-8   Small ellipse templates.

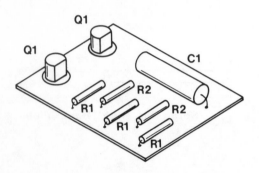

Figure 8-9   An isometric drawing of a circuit board.

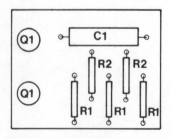

Figure 8-10   A pictorial schematic drawing from which Fig. 8-9 was created.

8-3   ISOMETRIC DRAWINGS OF CIRCUIT BOARDS

Figure 8-9 shows an isometric drawing of a circuit board. It is actually a composite of several smaller isometric drawings: 5 isometric drawings of resistors, 1 isometric drawing of a capacitor, 2 isometric drawings of transistors, and a circuit board, which are all combined to form the finished isometric drawing.

Figure 8-9 was drawn using the information presented by the pictorial schematic diagram shown in Figure 8-10. For example purposes the wiring paths have been omitted. The procedure used to create the isometric drawing shown in Figure 8-9 from the schematic diagram shown in Figure 8-11, is as follows:

1. Using very light lines, lay out an isometric drawing of the circuit board.

2. Locate the components on the circuit board as shown.

3. Draw isometrically each individual component.

4. Trace; from the layout created in step 3, the finished isometric drawing. Draftsmen prefer to trace the finished drawing, since the layout created in step 3 is usually a maze of construction lines which is impossible to erase cleanly.

To summarize the procedure and to make it easier for you to prepare isometric drawings, Figure 8-12 has been included. Figure 8-12 is a pictorial listing of the components most often called for when creating isometric drawings. In each case, the step-by-step procedure used to create the finished isometric drawing is shown.

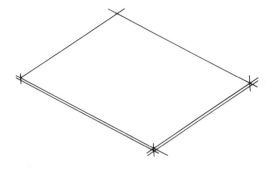

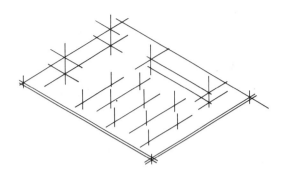

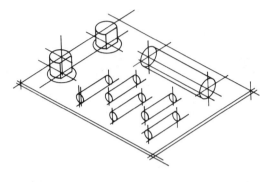

Figure 8-11 How to prepare an isometric drawing of a circuit board.

Circular Indicator

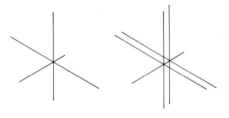
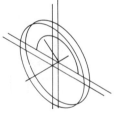

shading

Jack Receptacle

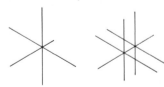

shading

Cylindrical Switch with Dial

shading

Figure 8-12 A pictorial listing of various components often used for isometric electronic drawings.

Cylindrical Switch with Collar

Throw Switch

Coil

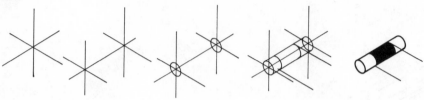

Diode

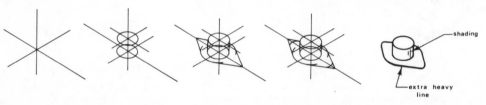

Meter

Cylindrical Switch

Tapered Switch

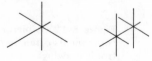

Figure 8-12 Continued.

Grooved Cylindrical Switch

Sliding Switch

Protruding Square Switch

Recessed Square Switch

Rectangular Indicator

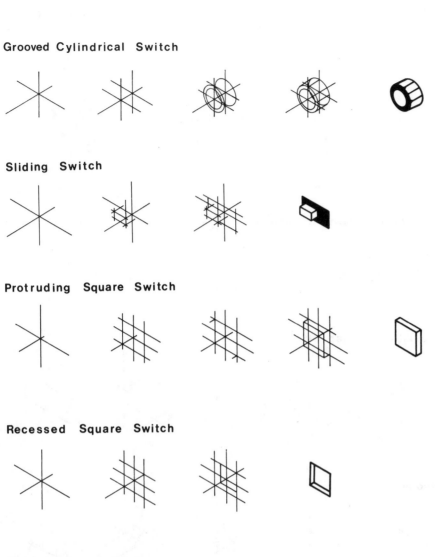

shading

Resistor or Fuse

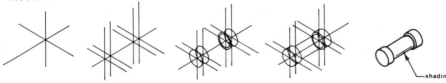

shading

Figure 8-12   Continued.

129

Transformer

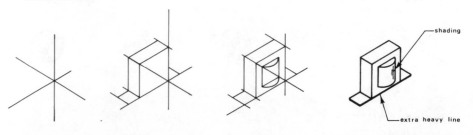

Capacitor (drawn freehand)

Resistor   or   Capacitor

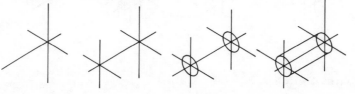

Diode

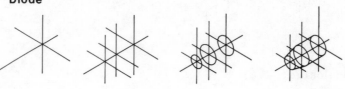

Transistor

Figure 8-12 Continued.

8-4 OBLIQUE DRAWINGS

Oblique drawings are pictorial drawings which are based on an oblique axis: that is, a vertical line, a horizontal line, and a receding line (usually 30°) (see Figure 8-13). The chief advantage in preparing

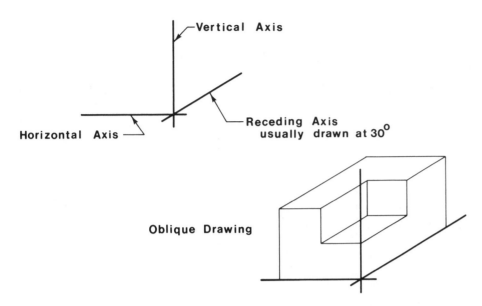

Figure 8-13 An oblique axis.

oblique drawings as opposed to other types of pictorial drawings is that the frontal plane contains a 90° angle. This means that round edges may be drawn as circles or arcs if they appear in the front plane or a plane parallel to the front plane. However, the advantage of being able to draw circles is applicable only to the frontal plane of oblique drawings, receding planes require elliptical shapes for visual correctness just as did isometric drawings. This means that unless the object can be positioned so that the majority of circles and arcs are located in the frontal plane or a plane parallel to the frontal plane, there is really no advantage to preparing an oblique drawing.

To prepare an oblique drawing, use the following procedure. The procedure is illustrated in Figure 8-14.

    1. Lay out an oblique axis.

    2. Draw the length, depth, and width of the object as shown.

    3. Add, using very light construction lines, the preliminary detail of the object.

    4. Add the final detail to the drawing.

    5. Erase any excess lines and draw, using heavy black lines, the final shape of the object. The final drawing may be traced if the layout created in step 4 can not be erased cleanly.

Oblique drawings tend to make objects look deeper than they actually are. For example, compare the isometric and oblique drawings of Figure 8-1. The oblique drawing appears to be larger, when in fact, both drawings are drawn to the same scale (check for yourself). To overcome this visual distortion, oblique drawings are often drawn as cabinet projections.

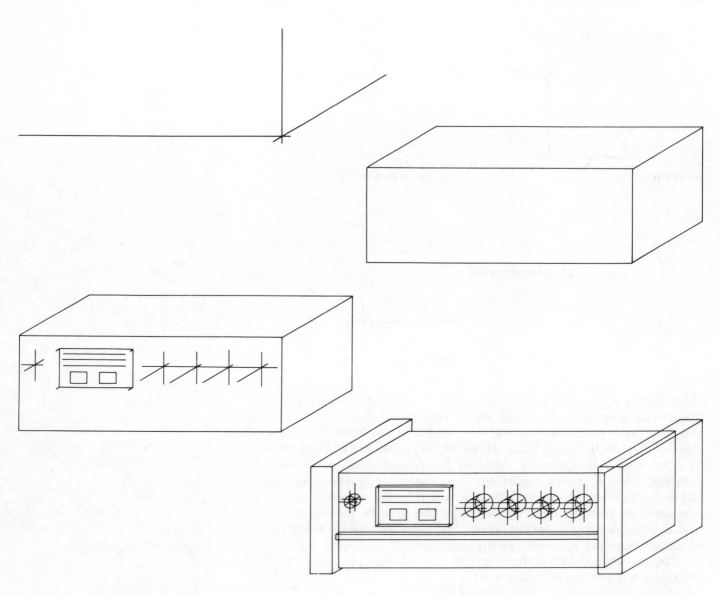

Figure 8-14 How to prepare an oblique drawing. The example used is a Heath kit stereo receiver AR–1515.

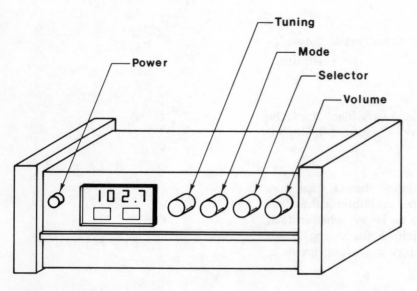

Power

Tuning

Mode

Selector

Volume

Cabinet projections are drawn exactly the same as regular oblique drawings (which are called cavalier projections) except dimensions along the receding axis are drawn at half size, that is, at $\frac{1}{2}$ their scale values. Figure 8-15 illustrates. Note, in Figure 8-15, how much more proportioned the cabinet projection appears relative to the isometric drawing than does the full-size oblique drawing. The only dimensional differences between the three cubes in Figure 8-15 is the half-scale reduction of the receding axis values in the cabinet projection. Take a scale or pair of dividers and compare the objects of Figure 8-15 for yourself.

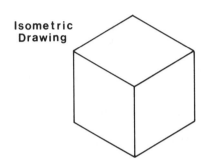

Isometric Drawing

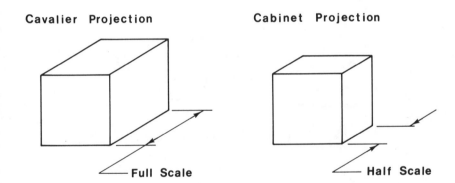

Cavalier Projection　　　　**Cabinet Projection**

Full Scale　　　　**Half Scale**

Figure 8-15 A comparison between a cavalier and cabinet projection.

8-5 OBLIQUE DRAWINGS OF CIRCUIT BOARDS

Oblique drawings may be made of circuit boards using the same procedure outlined in Section 8-3 for isometric drawings. The only difference is that when creating an oblique drawing an oblique axis must be used.

Figure 8-12, which is set up to help create isometric drawings, may be modified to help create oblique drawings by changing the first step in each drawing outline from an isometric axis to an oblique axis. Figure 8-16 illustrates this change as applied to the drawing of a diode and a meter with a recessed face. Figure 8-17 is an example of an oblique drawing of a circuit board.

Diode

Meter

Figure 8-16 Oblique drawings of components. The procedure used to create the drawings is the same as was used for Fig. 8-12. The only difference is the initial axis setup.

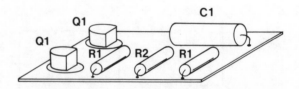

Figure 8-17 An oblique drawing of a circuit board.

PROBLEMS

8-1　Prepare either an isometric or oblique drawing, as assigned by your instructor, of the following realistic components (photographs are courtesy of Tandy Corp. Radio Shack). All dimensions are to be by eye, but try to keep the general proportions as pictured. The microphones and mounting brackets may be omitted.

　　a. Figure 8-18
　　b. Figure 8-19
　　c. Figure 8-20
　　d. Figure 8-21 (use a large sheet of paper)

Figure 8-18

Figure 8-19

Figure 8-20

Figure 8-21

8-2   Prepare an isometric drawing of the component side of the PC board shown in Figure 8-22.

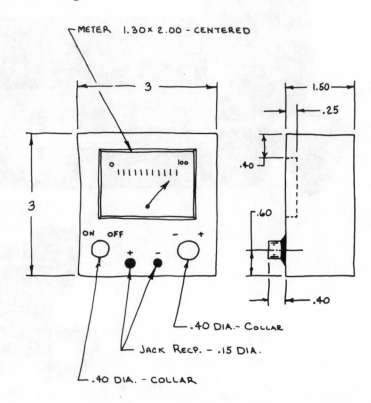

METER 1.30 × 2.00 - CENTERED

.40 DIA. - Collar

JACK RECP. - .15 DIA.

.40 DIA. - COLLAR

METER BOX

Figure 8-23

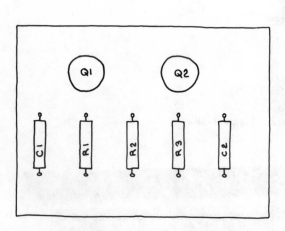

TRANSISTORS = .80 DIA. × .35 HIGH

RESISTORS & CAPACITORS = .30 DIA. × 1.00 LONG

BOARD SIZE = 4.00 × 5.00

Figure 8-22

8-3   Prepare an isometric drawing of the meter box sketched in Figure 8-23.

8-4   Prepare an isometric drawing of an amplifier which measures 4 × 12 × 12. Design the front panel so it includes an off-on pushbutton switch, a volume control, a tone switch, a speaker control switch labeled *A*, *B*, *A + B*, a balance control switch, a switch marker PHONO, TAPE, MIC, MISC, and an earphone jack plug.

GRAPHS
AND CHARTS

9-1 INTRODUCTION

This chapter explains how to create graphs and charts. Draftsmen are often asked to convert lists of raw data into graph or chart form so that the data may be more easily understood. Most engineering test data, for example, originally appears in computer printout form — long, long lists of numbers. To read and analyze these lists is tedious work. If however, the same data is presented in graph form, it is much easier to understand.

This chapter covers pie charts, bar charts, and curve plotting. Logarithmic and semilogarithmic grids and scales are also included.

9-2 PIE CHARTS

Pie charts are used to illustrate the relative sizes of various component parts of a total quantity. The name pie chart comes from its shape which looks like an overview of a pie which has been sliced into different-sized pieces. Figure 9–1 illustrates a pie chart.

To demonstrate how to create pie charts, consider the following problem (Figure 9-2 illustrates). You are asked to draw a pie chart which will illustrate the relative numbers of persons serving the Army, Navy, Air Force and Marine Corps in 1975. You are given the following figures:

1975 — Military Personnel

Army	781,316
Navy	549,400
Air Force	608,337
Marines	192,200

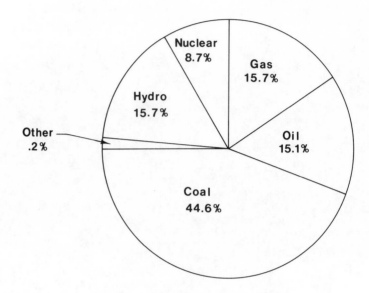

Figure 9-1   An example of a pie chart.

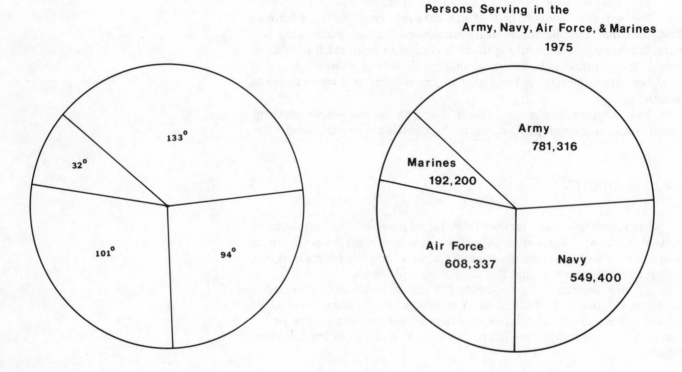

**Persons Serving in the
Army, Navy, Air Force, & Marines
1975**

Figure 9-2   How to prepare a pie chart. The degree values are
based on calculations.

To convert these figures into a pie chart:

1. Calculate the percentage value of each figure. This is done by first finding the total value of all the figures and then dividing each figure into the total.

$$
\begin{array}{l}
781,316 \\
549,400 \\
608,337 \\
\underline{192,200} \\
2,131,253 \quad \text{Total value}
\end{array}
$$

Army $\dfrac{781,316}{2,131,253} = 0.37$ or 37%

Navy $\dfrac{549,400}{2,131,253} = 0.26$ or 26%

Air Force $\dfrac{608,337}{2,131,253} = 0.28$ or 28%

Marines $\dfrac{192,200}{2,131,253} = 0.09$ or 9%

2. Convert the percentages calculated in step 1 into an equivalent percentage of a circle. This is done by multiplying the percentage values by $360°$.

Army 37% of $360° = 0.37\,(360°) = 133°$
Navy 26% of $360° = 0.26\,(360°) = 94°$
Air Force 28% of $360° = 0.28\,(360°) = 101°$
Marines 9% of $360° = 0.09\,(360°) = 32°$

These values are in terms of degrees, and can be measured using a protractor.

3. Lay out a circle about $6''$ in diameter and mark off the degree values calculated in step 2.

4. Label each sector as shown and add a title for the chart. All lettering should be neat and easy to read. If desired, the different sectors may be shaded differently to help distinguish them from each other.

9-3   BAR CHARTS

Bar charts are used to demonstrate the differences between fixed quantities. They derive their name from the fact that they express values in terms of bar-shaped figures as shown in Figure 9-3.

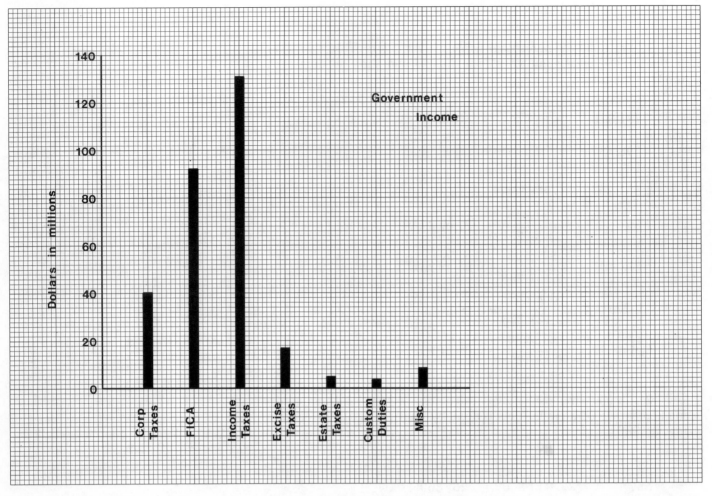

Figure 9-3 An example of a bar chart.

It is important that bar charts be drawn with a properly proportioned scale. Scales which are too small are difficult to read and may not show the differences between values clearly, while scales which are too large may not fit the paper.

Because bar charts usually involve a comparison of fixed values, the values are sometimes lettered in just above the top of the bar. In this way the viewer gets not only a visual comparison but also the numerical values so that further comparison may be made if desired. This information could be derived from the bar chart scale, but it is usually very helpful to the reader to have the values clearly stated.

Consider the following problem (Figure 9-4 illustrates). You are asked to create a bar chart which compares the number of home runs hit by the National League's leading home run hitters from 1920 to 1924. The number of home runs hit was:

1920	Cy Williams, Philadelphia	15
1921	George Kelly, New York	23
1922	Roger Hornsby, St. Louis	42
1923	Cy Williams, Philadelphia	41
1924	Jacques Fourmier, Brooklyn	27

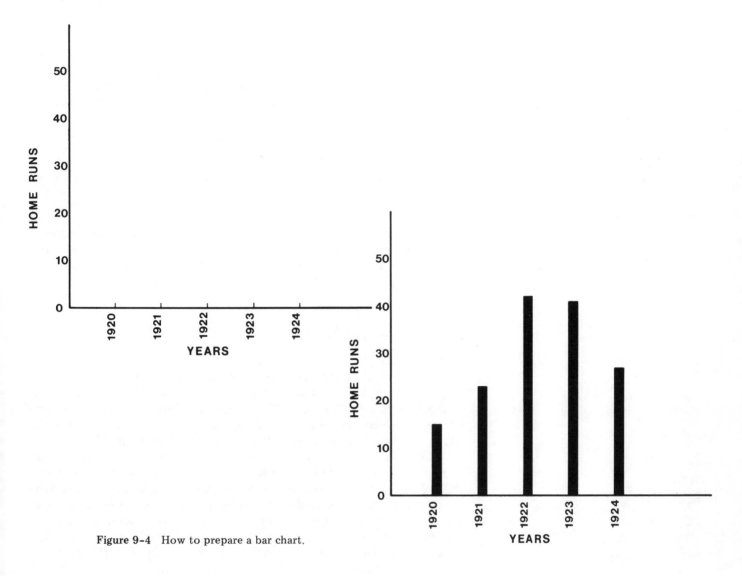

Figure 9-4   How to prepare a bar chart.

National League's
Home Run Leaders
1920 - 1924

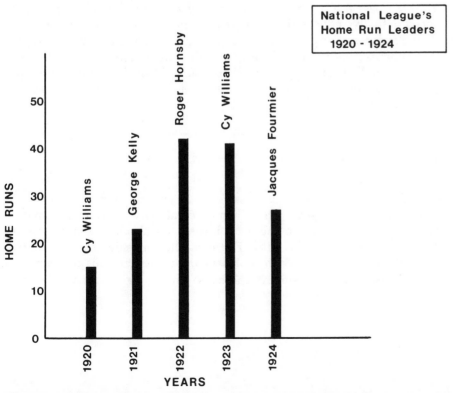

To convert these figures into a bar chart:

1. Choose a scale. In this case we let each block on the vertical axis represent 1 home run, or 10 home runs to the inch. Whenever possible, choose a scale which is directly related to 10 as most people are used to thinking in terms of a base 10.

2. Define the horizontal and vertical axis.

3. Draw in the bars.

4. Label each bar and add the title and any other necessary information.

It is not necessary to have the horizontal axis represent a value of 0. If there are extremely large values to work with, the length of the bars may be reduced by assigning a value higher than 0 to the horizontal axis.

9-4   CURVE PLOTTING

Curve plotting is a term which is used to describe the creation of graphs that are based on data which is a function of two or more variables (Figure 9-5 illustrates). Curve plotting does not produce a comparison between fixed values as do pie charts and bar charts, but instead shows the relationship between two or more different variables.

For example, in Figure 9-5, we see the horizontal axis is a measure of engine speed (RPM) and the vertical axis is a measure of horsepower (HP). Both engine speed and horsepower are variables. The shape of the curve shows the relationship between the two variables. The curve slopes upwards meaning as values of engine speed increase, values of horsepower also increase up to the point of maximum horsepower. After this point as the values of engine speed increases, the values of horsepower decrease — the curve slopes downwards. This means that after we pass the point of maximum horsepower, increasing the engine speed actually reduces the amount of horsepower the engine generates.

Four areas often cause trouble when plotting curves; choosing a scale, identifying the curves properly, drawing a smooth curve between the data points, and choosing the proper grid background (sometimes patterns other than squares are used). Each area will be considered separately.

9-5   CHOOSING A SCALE

A scale must be chosen so that it will be easy to read. It is tempting to pick a scale which is mathematically convenient — that is, the scale is easy to fit to the data, but such scales are usually very difficult to read and work with. For example, consider the two curves plotted in Figure 9-6. Both represent the same data but the curve on the left uses an odd scale and the curve on the right uses a scale based

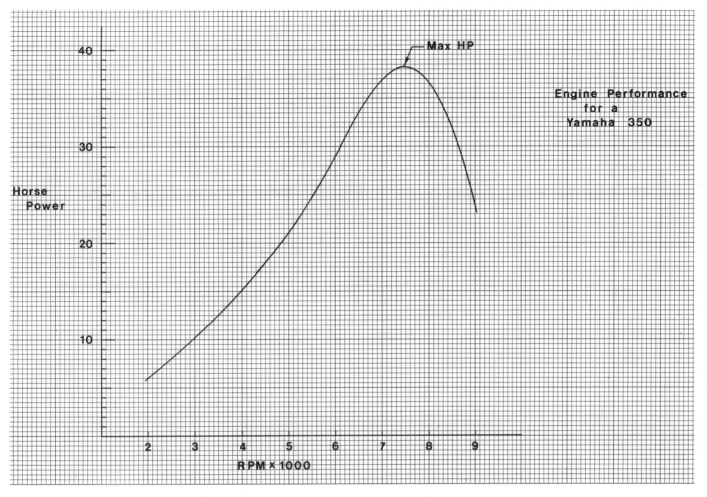

Figure 9-5 An example of a curve which shows the relationship between two or more variables, in this case horsepower vs. rpm.

Figure 9-6 Care must be taken when choosing a scale. What is the value of Point A?

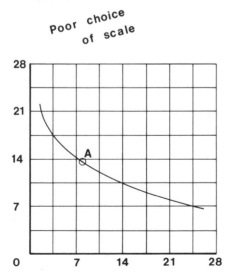

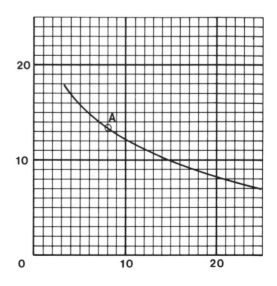

on 10. What is the x and y value of point A? If we use the curve on the right we can easily find the answer $x = 8.0$, $y = 13.3$, but we see how much more difficult it is do derive the same answer from the left curve.

A scale must also consider the size of the paper the curves are to be drawn on. Most $8^{1}/_{2}$ by 11 sheets of graph paper have a grid pattern which is 8 by 10 (some types of graph paper have even smaller grids). The workable area of the paper is further reduced when the space needed to define the horizontal and vertical axis is taken into account. This means that the final workable area—the area in which we can draw curves—for an $8^{1}/_{2}$ by 11 sheet of paper is only about 7 by 9 at most (see Figure 9–7).

To choose a scale which is both easy to read and will fit the paper, consider the following problem.

Plot the performance curve for a Texas Instruments SN52506 Dual Differential Comparator with Strobe as a function of input offset current versus the free air temperature. The test data is

Free air temperature (°C)	Input Offset Current (μA)
–50	2.09
–25	1.40
0	0.94
25	0.67
50	0.53
75	0.43
100	0.36

We first notice that the free air values are equally spaced and that at least 6 spaces are needed. We also realize that the free air tempera-

Figure 9–7 Always leave room for identifying the axis. This means that the area for curve plotting is actually less than the total grid pattern.

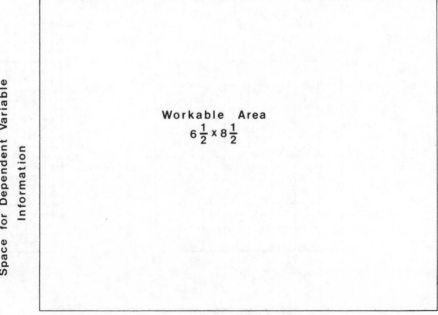

Space for Dependent Variable Information

Workable Area
$6\frac{1}{2} \times 8\frac{1}{2}$

Space for Independent Variable
Information

ture values are the independent variables (the temperature of air does not depend on the values of the comparator) and should be plotted along the horizontal axis.

If we let each inch represent 25° of free air temperature, we need at least 6 inches along the horizontal axis. 25 units per inch would not normally be a good choice but because we have evenly spaced values it is acceptable in this situation.

The input offset current values vary from 0.36 to 2.09. If we let each inch equal 1.00 μA we would need a little over 2 inches along the vertical axis. However, this would result in an almost flat curve which would not clearly represent the relationship between the variables. By doubling the scale so that 2 blocks equals 0.1 μA the curve will require a little over 4 inches on the vertical scale.

It is customary to extend curves beyond their furthest data points so we need, using the scales determined above, approximately a 5 × 7 area for the curve. This is well within the 6½ × 8½ limits outlined in Figure 9-7. Figure 9-8 illustrates how the final curve was plotted and labeled.

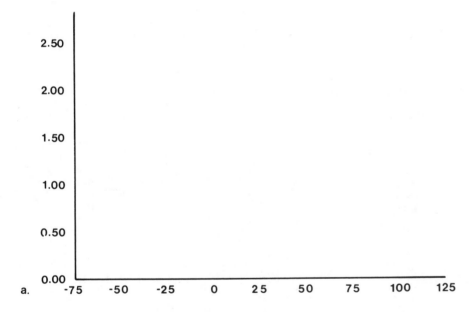

Figure 9-8 How to prepare a curve plot: (a) define axis and label; (b) add data points and draw in curve; (c) add all required labels.

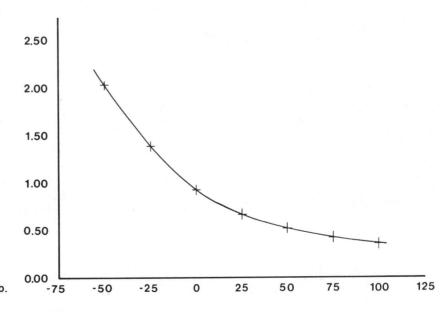

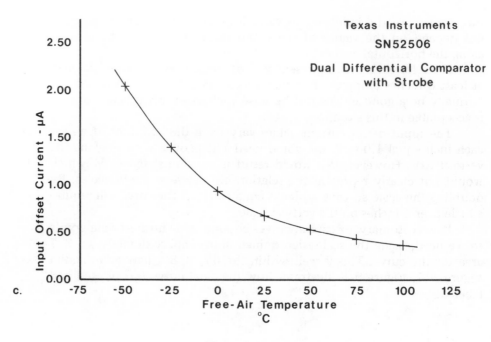

Texas Instruments
SN52506

Dual Differential Comparator
with Strobe

Figure 9-8 Continued.

Figure 9-9 A sample curve plot of Age vs. Height for males and females.

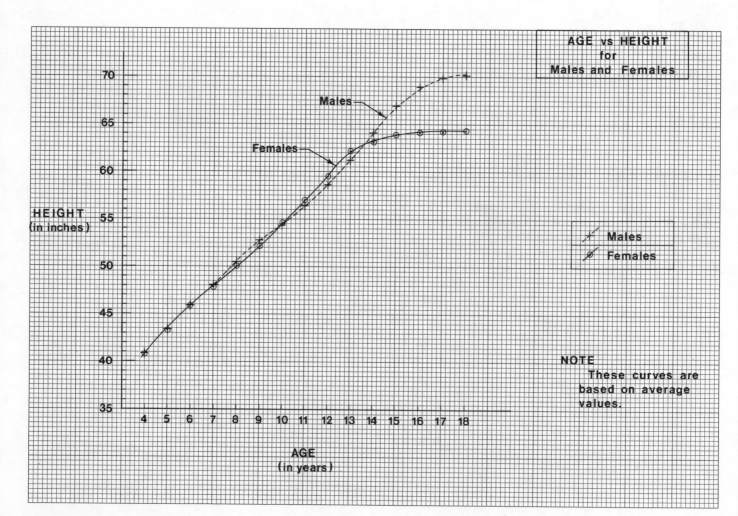

Figure 9-9 plots age versus height for males and females. Because the curves are almost identical up until age 12, a large vertical scale is required to bring out the small differences. Note how small lines were added to both the horizontal and vertical axis to make it easier to understand the scale used.

Note how in all curve plots the final graph includes completely defined horizontal and vertical scales including labels, a title, a definition of data points and line patterns, any notes required, and labels for all curves.

9-6 IDENTIFYING CURVES

When two or more curves are drawn on the same graph, it is important that each be clearly identified. Several drawing techniques may be used to distinguish curves. A different color may be used for each curve, but most reproduction processes available to draftsmen do not reproduce color, so other methods must be used.

Figure 9-10 illustrates four different line patterns, and four different data point symbols, which can be used together to help distinguish curves. Note how curve 1 uses a small circle to indicate its data points and a broken line pattern for the actual curve. Each of the other curves uses a different line pattern and data point symbol, all of which are defined in the accompanying legend. Each curve may be further labeled as shown but this is an optional practice.

Figure 9-10 When plotting different curves on the same axis, use different line patterns and different type data point symbols to help the reader distinguish the curves.

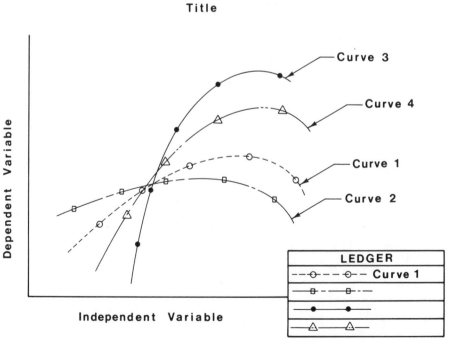

Once data points have been located on the graph paper, the problem still remains of joining them with a smooth curve. Draftsmen create smooth curves by using plastic curves or adjustable curves as guides. Some examples of these devices are pictured in Figure 9–11.

Most students make the error of trying to connect too many points with one positioning of the curve. Figure 9–12 shows a series of points that are partially connected. The curve is in position to serve as a guide for joining *only* points 3 and 4 — not 3, 4, and 5 — even though all three seem to be aligned. To join point 5 using the curve position as shown would make it almost impossible to draw a smooth curve.

It is sometimes *not* necessary to connect all data points with a curve which goes through every data point. Curves which pass through every data point may be so wavy that they would be almost impossible

Figure 9–11 Some of the many different types of curves used by draftsmen to draw curves.

Figure 9–12 How to draw a smooth curve between data points.

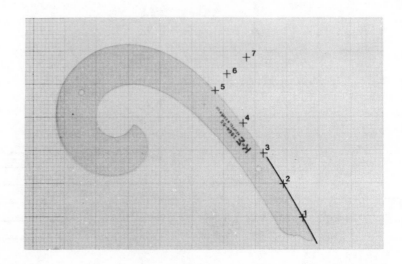

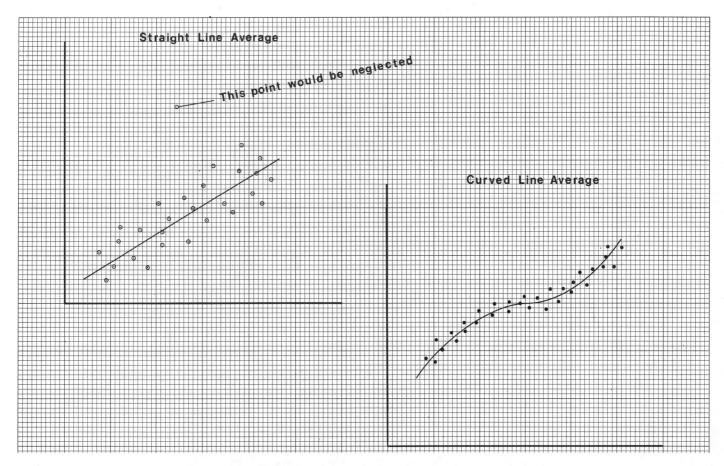

Figure 9-13 Examples of a straight line average and a curved line average.

to analyze. Further, in many cases, an exact analysis is not necessary since basic trends and patterns can be derived from an *average curve*.

Figure 9-13 shows two different types of average curves. The straight line average is created by drawing a straight line through the data points so that approximately half the points are above the line and half below the line. The distance between the points above the line should be approximately equal to the distance between the line and the points below the line. Curved line approximations can be made in a similar manner.

9-8 CHOOSING A GRID PATTERN

Up to now, we have considered only square grids, that is, patterns whose lines are evenly spaced in both the horizontal and vertical directions. Many other patterns are used to plot data, including logarithmic, semilogarithmic, and many different circular grids. Figure 9-14 pictures a sheet of graph paper printed with a polar grid pattern.

Figure 9-15 illustrates a logarithmic and semilogarithmic grid pattern. The logarithmic pattern is usually referred to as log-log paper and the semilogarithmic pattern as semilog paper. Both grids are based on the logarithmic scale, also illustrated in Figure 9-15.

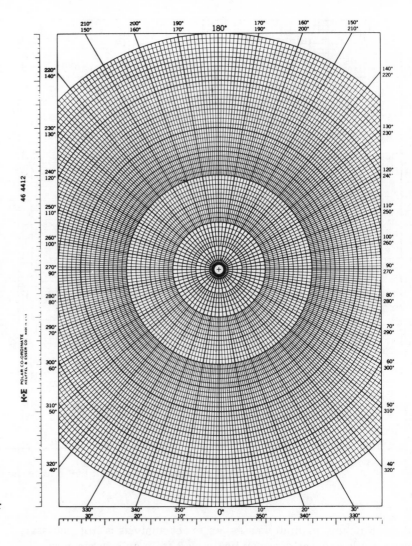

Figure 9-14 A sample sheet of polar graph paper.

Figure 9-15 Samples of a logarithmic and semilogarithmic grid pattern and a sample logarithmic scale.

Logarithmic Grid (log–log)

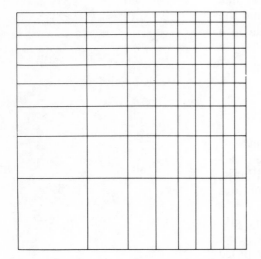

Semilogarithmic Grid (semi-log)

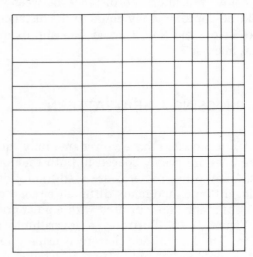

Logarithmic Scale

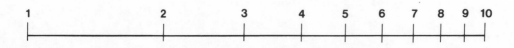

Log-log and semilog grids are used because, for some data, they make analysis easier. After draftsmen plot up a series of data, engineers usually try to analyze it by trying to find the mathematical equation that matches the curve. The work is, of course, greatly simplified if the data plots as a straight line, because the equation for any straight line is:

$$y = mx + c$$

Some data which plots as a curve on a square grid pattern plots a straight line on a logarithmic grid, as Figure 9-16 illustrates.

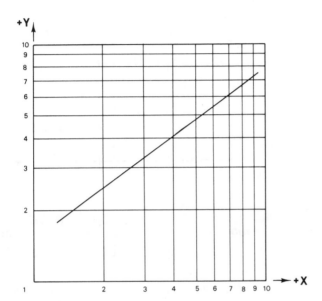

X	Y
2	2.52
3	3.36
4	4.10
5	4.75
6	5.46
7	6.08
8	6.62

Figure 9-16   A sample of data points which plot as a straight line on a log-log pattern.

The basic logarithmic grid can represent any multiple of 10 of the numbers 1 through 10. It can represent 1 through 10, 10 through 100, 100 through 1000, 0.1 through 1.0, etc. For example, point *A* in Figure 9-17 can equal 0.034, 0.34, 3.40, 34 or 340 depending on which scale is used. Likewise point *B* can equal 0.062, 0.62, 6.20, 62, or 620.

A unique feature of log grids is that they can *never* have a line with a value equal to 0.0. A line may be defined as 0.00001 or even smaller, but it can never be 0.0. This means that curves drawn on log grids can never cross, or even touch, the vertical axis.

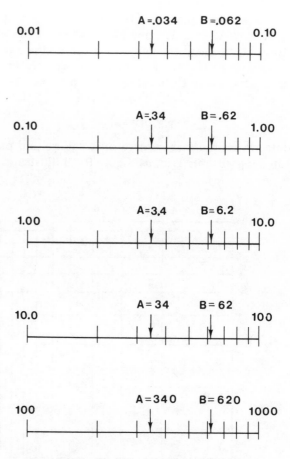

Figure 9-17 How to read different log scales.

Log scales may be placed one after another as shown in Figure 9-16 to increase the overall value range of the graph. It is good practice to label at least, the beginning and end values of each log pattern used.

PROBLEMS

9-1   Figure 9-18 illustrates three different mathematical curves; that is, curves whose data points were obtained by substituting values into mathematical equations. In each example the x variables are considered the independent variable and the y variables are considered the dependent.

Plot the following equations as assigned by your instructor. Include your calculation sheets. Always define the x variables as the independent variables.

   a. $y = x$
   b. $y = 2x$
   c. $y = -\frac{1}{2} x$
   d. $y = 1.6x + 2$

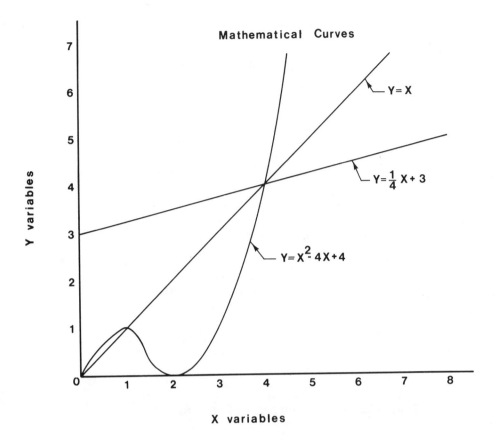

Mathematical Curves

Y = X

$Y = \frac{1}{4} X + 3$

$Y = X^2 - 4X + 4$

Y variables

X variables

Figure 9-18

e. $4y = 2x - 1.50$
f. $y = x^2$
g. $y = x^2 + 2x$
h. $y = \frac{1}{2} x^2 + 2x + 4$
i. $y = 1.6x^2 - 2.2x - 0.7$
j. $2y = -x^2 + 3x - 0.5$
k. $xy = 4$

9-2 Draw a pie chart based on the following data:

World Power Consumption, 1976	
Country	Kilowatt Hours
U.S.A.	1,999,688
U.S.S.R.	1,038,000
Japan	432,757
West Germany	301,800
Others	2,480,755
World total	6,253,000

9-3   Draw a pie chart based on the following data:

U.S. Government Crime Report, 1977

Murder	20,510
Forcible rape	56,090
Robbery	464,970
Aggravated assault	484,710
Burglary	3,252,100
Larceny-theft	5,977,700
Auto theft	1,000,000
Total	11,256,000

9-4   Draw a bar chart based on the following data:

Electricity Generated by Atomic Power

Country	Output in Megawatts
U.S.A.	28,236
Britain	4,258
U.S.S.R.	4,521
Japan	3,718
France	2,886

9-5   Draw a bar chart based on the following data:

Leading Lifetime Scores Thru 1977
Professional Football

Name	Points
George Blanda	2,002
Lou Groza	1,608
Fred Cox	1,227
Jim Bakken	1,171
Jim Turner	1,147
Gino Cappelletti	1,130
Bruce Gossett	1,031

9-6   Draw a bar chart based on the following data:

Leading Rebounders in NBA – 1975–1976

Name	Team	Rebounds
Abdul-Jabbar	Los Angeles	1,383
Cowens	Boston	1,246
Unseld	Washington	1,036
Silas	Boston	1,025
Lacey	Kansas City	1,024

9-7 Plot the following data. Plot the years along the horizontal axis and the hourly earnings along the vertical axis.

Increases in Factory Workers
Hourly Earnings

Year	Hourly Earnings
1965	2.61
1966	2.72
1967	2.83
1968	3.01
1969	3.19
1970	3.36
1971	3.56
1972	3.81

9-8 Ohm's law defines the relationship between current, voltage and resistance as

$$I = \frac{E}{R}$$

where

I = Current (amperes)

E = Voltage (volts)

R = Resistance (ohms)

Plot a family of curves comparing amperes and volts at four different values of ohms. Plot the volts along the horizontal axis.

The values to be plotted may be calculated by first assuming a constant value for ohms and then inserting various values for volts and solving the equation for amperes. For example, assume the ohms value equals 2 ohms. If we solve the equations for voltage values of 1, 2, 3, 4, and 5, we have

$$I = \frac{1}{2} = 0.5$$

$$I = \frac{2}{2} = 1.0$$

$$I = \frac{3}{2} = 1.5$$

$$I = \frac{4}{2} = 2.0$$

$$I = \frac{5}{2} = 2.5$$

We can then plot the values

Volts	Amperes
1	0.5
2	1.0
3	1.5
4	2.0
5	2.5

Label each curve as to its constant ohms value.

9-9   Plot on semilog paper the following equations.

    a. $y = x$
    b. $y = 3x + 2$
    c. $y = x^2$
    d. $y = x^2 + 1.5x - 4$
    e. $y = e^x$
    f. $y = e^{x+2}$
    g. $y = e^x + 0.60$

9-10   Plot on log-log paper the following equations.

    a. $y = x$
    b. $y = -x^2 + 2.4$
    c. $y = e^x$
    d. $y = e^{x+2}$
    e. $y = e^x + 2$

RESIDENTIAL ELECTRICAL WIRING

10-1 INTRODUCTION

This chapter deals with residential electrical wiring drawings. The chapter will explain basic fundamentals of architectural drawing, define and show the application of the graphic symbols used on residential wiring drawings, and cover some of the basic design concepts required for residential wiring.

The information presented is in agreement with the standards set forth in the National Electrical Code published by the National Fire Protection Association, 470 Atlantic Ave., Boston, MA 02110. Interested students may purchase copies of the Code by writing to the NFPA at the above address.

10-2 BASIC ARCHITECTURAL DRAFTING

The principal type of drawing used by architects is called a floor plan. A floor plan, as the name implies, is a drawing which defines the location of the various rooms, windows, doors, stairs, closets, hallways, etc. for a given residence. Figure 10-1 shows a floor plan along with an elevation drawing of a small one bedroom house.

The floor plan in Figure 10-1 has been labeled so that you can learn how to interpret architectural drawings. Note how doors, windows, etc., are represented. All linework is as outlined in Figure 1-17 and all lettering is basically the same as shown in Figure 1-9 although it should be pointed out that architectural lettering is usually much more stylish than basic mechanical lettering.

The standard scale used for residential drawings is $\frac{1}{4}'' = 1'$; every $\frac{1}{4}''$ on the drawing equals $1'$ on the house. Figure 10-2 illustrates a $\frac{1}{4}''$ scale along with some sample measurements. The $\frac{1}{4}''$ to the right of the

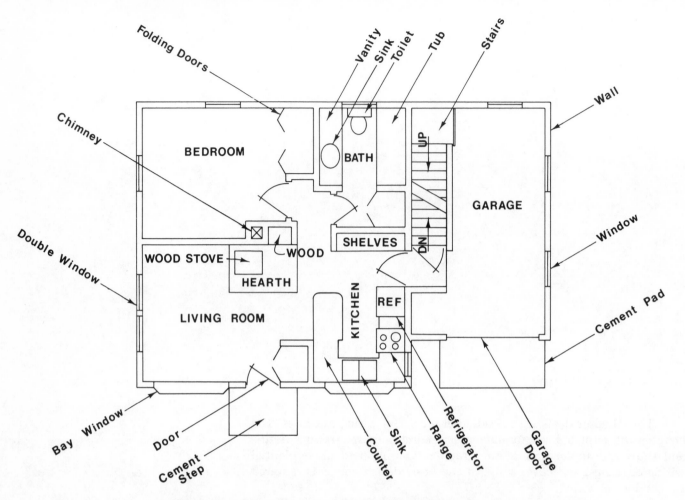

Folding Doors
Vanity
Sink
Toilet
Tub
Stairs
Wall
Chimney
Window
Double Window
Cement Pad
Bay Window
Door
Cement Step
Counter
Sink
Range
Refrigerator
Garage Door

BEDROOM
BATH
GARAGE
UP
WOOD STOVE
WOOD
SHELVES
DN
HEARTH
REF
LIVING ROOM
KITCHEN

Figure 10-1 (a) & (b) A floor plan and elevation drawing of a small house. The elevation drawing was prepared by Prof. Raymond Collard and the house design was created by Prof. Arthur Nelson both of Wentworth Institute of Technology.

Readings are based on a scale of $\frac{1''}{4}=1'$

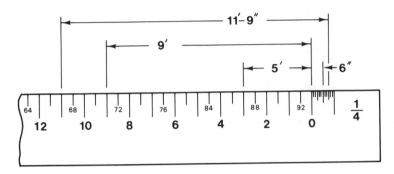

Figure 10-2 A ¼″ scale with some sample readings.

0 mark on the sample scale is graduated into 12 equal divisions so that measurements in inches can be made. Some scale manufacturers combine the ¼″ scale, which reads from right to left, with a ⅛″ scale, which reads left to right. The smaller numbers included in Figure 10-2 refer to values on a ⅛″ scale.

10-3 ELECTRICAL SYMBOLS

Symbols are used on residential electrical wiring drawings to specify the type and location of the switches or outlets required. Figure 10-3 illustrates the most commonly used symbols and their meaning. The dimensions given are only approximations as no national standard sizes have been agreed upon.

Figure 10-4 illustrates how these symbols may be added to a floor plan to create an electrical wiring drawing. In some cases, the electrical wiring drawing is created by making a blueprint of the floor plan and then drawing the appropriate symbols on with either a red or yellow pencil.

Wiring paths are drawn as freehand hidden lines (a series of dashes, see Figure 1-17). They are drawn freehand so that they are easily distinguishable from the floor plan lines. Note that the wire paths are only specified between switches and the fixtures they activate. It is not necessary to show all the wiring paths within a residence.

Wiring paths are omitted for two reasons. To show all the wiring would create a very cluttered, difficult-to-follow drawing which could lead to interpretation errors. It is also not necessary to show all the paths. Electricians need only be shown the location and type of fixture required and they will be responsible to insure that the fixture is wired properly in accordance with the national code and any local variances.

If you have trouble drawing neat freehand lines, use a french curve as a guide. The curve should help steady your hand.

ELECTRICAL SYMBOLS

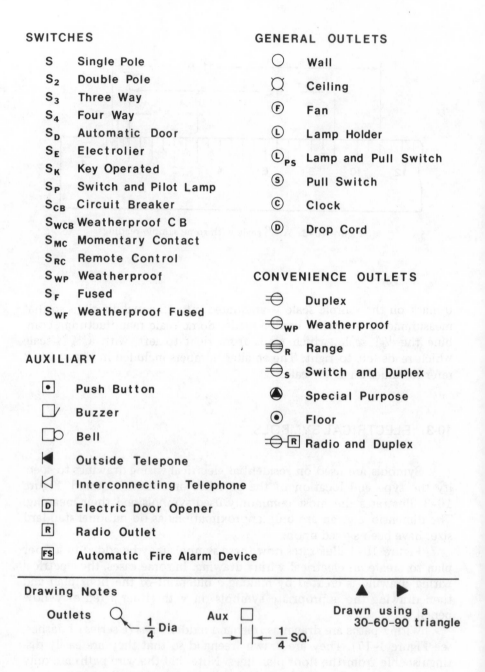

SWITCHES

S	Single Pole
S_2	Double Pole
S_3	Three Way
S_4	Four Way
S_D	Automatic Door
S_E	Electrolier
S_K	Key Operated
S_P	Switch and Pilot Lamp
S_{CB}	Circuit Breaker
S_{WCB}	Weatherproof C B
S_{MC}	Momentary Contact
S_{RC}	Remote Control
S_{WP}	Weatherproof
S_F	Fused
S_{WF}	Weatherproof Fused

AUXILIARY

- Push Button
- Buzzer
- Bell
- Outside Telepone
- Interconnecting Telephone
- Electric Door Opener
- Radio Outlet
- Automatic Fire Alarm Device

GENERAL OUTLETS

- Wall
- Ceiling
- Fan
- Lamp Holder
- Lamp and Pull Switch
- Pull Switch
- Clock
- Drop Cord

CONVENIENCE OUTLETS

- Duplex
- Weatherproof
- Range
- Switch and Duplex
- Special Purpose
- Floor
- Radio and Duplex

Drawing Notes

Outlets ¼ Dia   Aux ¼ SQ.   Drawn using a 30-60-90 triangle

Figure 10-3 Electrical symbols for residential wiring drawings.

Every residential electrical wiring drawing you create, should include a symbol ledger which illustrates each symbol used on the drawing and defines its meaning. A sample symbol ledger is located on the drawings presented in Figures 10-4. The ledger may also be used to define any new symbols needed. Sometimes special equipment, such as smoke detectors, must be specified on a drawing and no standard

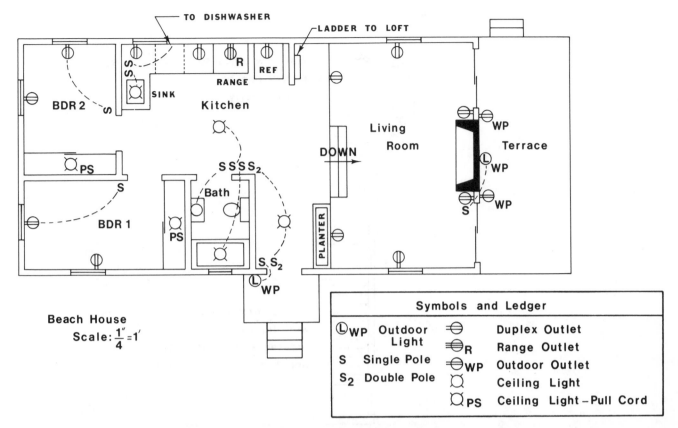

Figure 10-4 An example of a residential wiring drawing. The floor plan is for a summer beach house.

symbol has been defined to cover the equipment. In such cases, a draftsman may make up a symbol *provided* it is clearly defined in the drawing ledger.

If a special or specific type of switch or outlet is required, it must also be defined on the drawing. This is usually done by printing the manufacturer's name and the part number of the fixture next to the appropriate symbol on the drawing. It may also be done by creating a special symbol and defining its meaning in the drawing ledger.

10-4 LOCATING SWITCHES AND OUTLETS

When choosing locations for electrical switches and outlets, it is important to consider how the switches and outlets will be used. For example, the switch shown in Figure 10-5 is not located properly. It is behind the door and would require a person entering the room to start to close the door in order to reach the switch. This is unacceptable, particularly if the room is dark.

Another error illustrated in Figure 10-5 concerns the stairs. There is no light located near or over the stairs, meaning the stairs cannot be seen clearly at night.

The kitchen, shown in Figure 10-5 is also not lighted properly, even though the large flourescent would seem to be able to generate more than enough light. The problem, is shadows. A person working at

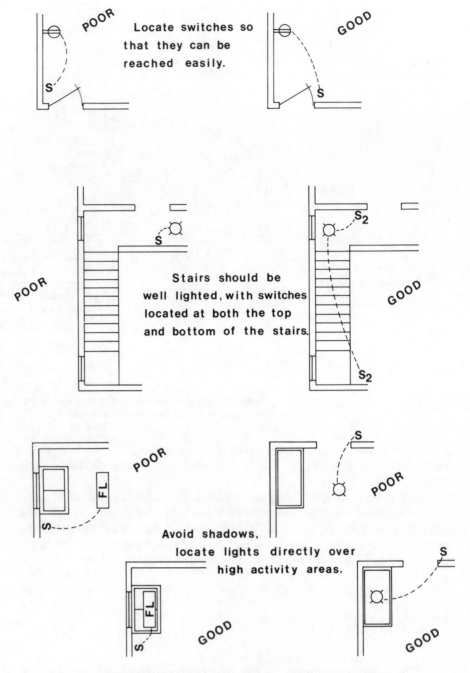

Figure 10-5 Examples of good and poor electrical design.

the sink would have the light at their back, causing a shadow to fall over the sink. Even though the room is well lighted, the sink would be in darkness making it inconvenient to use.

Shadows can also cause problems in the bathroom. In Figure 10-5 there is a light over the sink, but none over the tub. If a shower curtain is added (would not necessarily appear on the floor plan) with the shown lighting setup, the light would be blocked out by the curtain and the person showering would be in greatly reduced light.

It is good practice, when preparing your drawing, to pretend you are actually in the house you are working on. Assume it is dark, and mentally walk through the house checking that all switches and outlets you have specified are conveniently located. For example, if you are working on a two-story house, and you are on the first floor, you will want a light over the stairs, a switch on the first floor to activate the light, and another switch at the top of the stairs to deactivate the light.

Lights can be arranged with two, three, or four separate activating switches. The symbols for double-pole, three-way, and four-way switches are shown in Figure 10-6.

To help you locate switches properly, the following checklist has been prepared. Check your drawings against it to help prevent embarrassing errors.

1. When entering a room, is a light switch easily accessible?

2. For rooms with two or more entrances, are there light switches located at each entrance?

3. Are all outside entrances lighted?

4. Are all hallways and stairs lighted with switches located at each end?

5. Are all work areas (sinks, counters, etc.) lighted so as to prevent excessive shadows?

6. Are all special electrical requirements (range, dishwasher, dryer, air conditioners, etc) taken into account?

7. Are all outside fixtures waterproof?

Figure 10-6 How to indicate double pole, three-way, and four-way switches on a drawing.

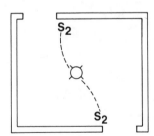

Two double pole switches wired to a single ceiling outlet; either switch can activate or deactivate the outlet.

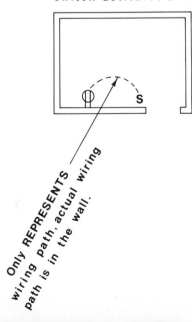

A single pole switch which is wired to a duplex outlet; the switch activates or deactivates the outlet.

Only REPRESENTS wiring path, actual wiring path is in the wall.

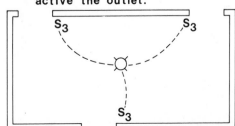

Three three-way switches all wired to a single ceiling outlet; any of three switches can be used to activate or deactive the outlet.

Several terms commonly used in residential electrical wiring are listed below. Each is defined and explained. Figure 10-7 illustrates some of the terms.

Branch Circuit. A power conducting wiring path that leads from the service panel to specific area of the house and back to the service panel. Some pieces of equipment used in residences draw so much electricity that a special individual branch circuit is created just for them. Electric ranges, dryers, refrigerators, air conditioners, and heaters are some of the equipment that would require an individual branch circuit.

Cable. Electrical wiring which has been covered with a protective cover or insulating material. Romex cable, copper wires covered with heavy paper and material for use in dry locations and plastic covered for damp locations, is the type most often used for houses. Armored cable, BX, is sometimes used.

Conduit. A thin-gaged metal pipe or tube which is used to protect cable from damage.

Drip Loop. A way to suspend the service wires so as to prevent water from running down the wire into the meter or service panel.

Mast. Conduit which extends above the roof of a house and is used to receive the service wires. Masts are not always used.

Meter Box. A gage installed by the local power company to keep track of the amount of electricity used.

Service Head. A metal cover placed on top of conduit to prevent water from entering the meter or service panel.

Figure 10-7 Residential wiring terms.

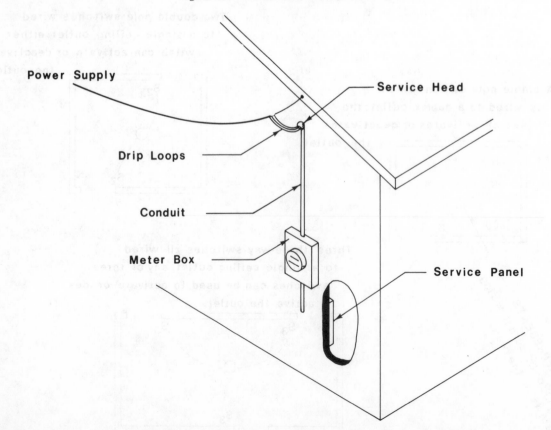

Service Panel. A metal box which contains the main disconnect switch and circuit breakers. It also contains all connection between the main incoming power and the branch circuits.

Service Wires. The power lines which bring electricity from the power company's lines to the house. The power supplied is usually 240/120-volt single-phase 60 cps AC.

15-amp Line, 20-amp Line, 240-volt Line. These terms refer to different types of branch circuits. A 15-amp line is an average line that would service a bedroom, bathroom, etc. A 20-amp line would be used for circuits which will probably get heavy use, such as a kitchen where many different appliances are used. A 240-volt line is a special, very heavy-duty branch circuit which is designed to carry large amounts of current. An electric hot water heater, electric range, electric dryer, etc., would all require 240-volt lines.

Figure 10-8 Floor plan of a bedroom.

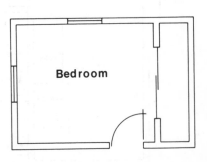

PROBLEMS

10-1 Figure 10-8 shows a floor plan of a bedroom, including a closet. Redraw the floor plan using a scale of $\frac{1}{4}'' = 1'$ and add, using the appropriate electrical symbols; 3 duplex outlets one of which is wired to a switch, a pull switch in the closet, and a telephone jack.

Each square of the background grid used in Figure 10-8 represents $\frac{1}{8}''$ per side.

10-2 Figure 10-9 shows a floor plan of a kitchen. Redraw the floor plan using a scale of $\frac{1}{4}'' = 1'$ and add, using the appropriate electrical sumbols, a

Range outlet
Dishwasher outlet
Refrigerator outlet
Overhead light with activating switches at each door
Light over the sink with an activating switch
Light and fan over the range with an activating switch for each
Four duplex outlets located over the counters
Wall outlet for clock

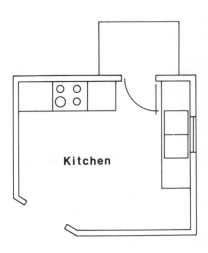

Each square of the background grid used in Figure 10-9 represents $\frac{1}{8}''$ per side.

10-3 Figure 10-10 shows the floor plans for a two-story house. Redraw the floor plans using a scale of $\frac{1}{4}'' = 1'$, and use the plans as a basis to create a set of residential electrical drawings. Think carefully about the locations of all switches, outlets, etc. before you add them to the drawings.

Each square of the background grid used in Figure 10-10 represents $\frac{1}{8}''$ per side.

Redraw the floor plans, as assigned by your instructor, as follows:

Figure 10-9 Floor plan of a kitchen.

a. First floor
b. Second floor
c. Foundation

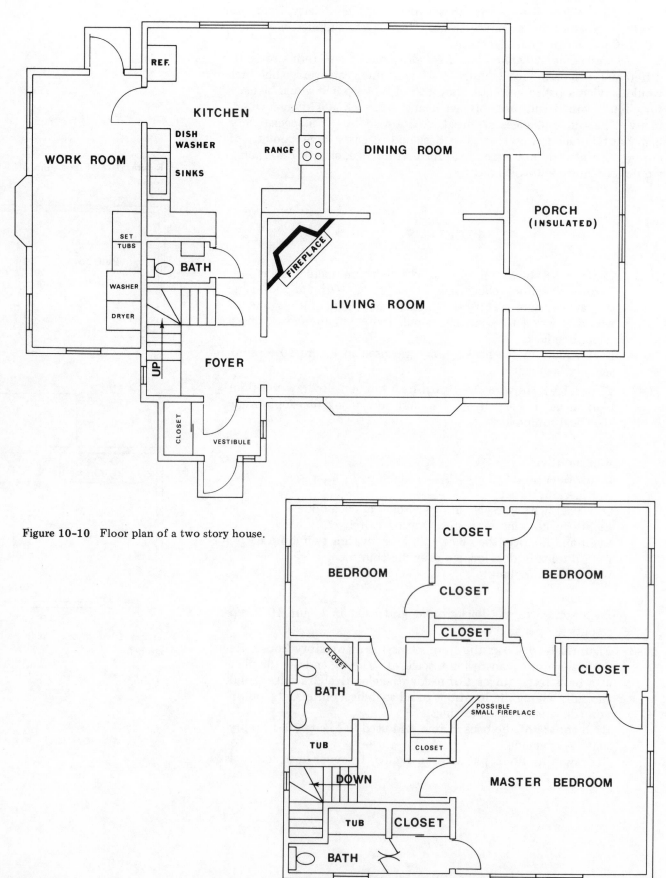

Figure 10-10 Floor plan of a two story house.

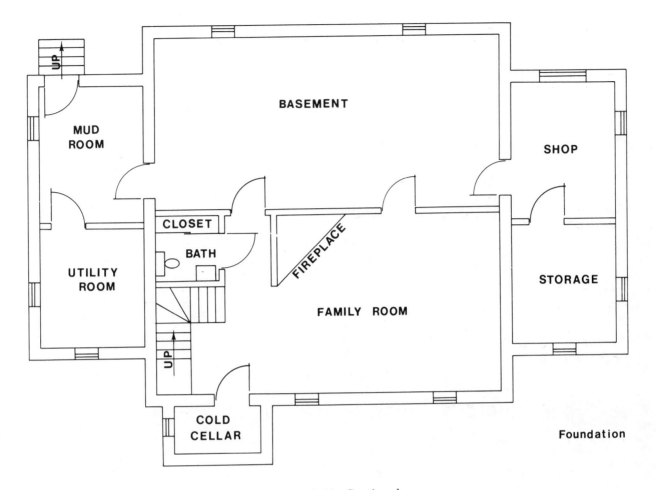

Foundation

Figure 10-10 Continued.

INDUSTRIAL
WIRING
DIAGRAMS

11-1   INTRODUCTION

This chapter explains how to draw some of the most widely used types of industrial wiring diagrams including: one-line diagrams, ladder diagrams, and raiser diagrams. The chapter will also present some of the basic design and electrical principles which these diagrams represent. The student interested in a more complete coverage of the subject is referred to:

> Charles W. Snow, *ELECTRICAL DRAFTING AND DESIGN*, 1976, Englewood Cliffs, N.J.: Prentice-Hall, Inc.

The symbols and line techniques required for each type of diagram are presented along with the text material. In addition, all the symbols are presented, for easy reference, in Appendix D.

11-2   BASIC POWER SYSTEMS

The design of any electrical system must first start with the power source. What kind and how much power is available? For example, most houses receive from the local power company power which is single-phase, three-wired, 120/240 volts, at 60 cps. This would be expressed on a drawing as:

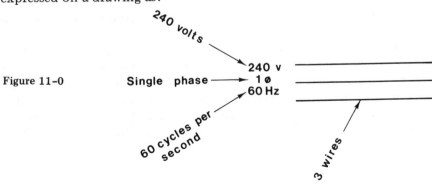

Figure 11–0

Three-Phase, Four-Wire Wye System

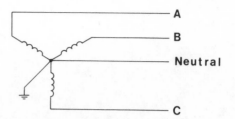

Used for apartment
buildings, office buildings,
shopping centers, etc.

Three-Phase, Three-Wire Delta System

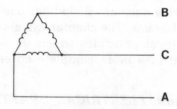

Used primarily for
factories or manufact-
uring shops.

Figure 11-1 A three-phase, four-wire Wye System and a three-phase, three-wire Delta System.

If more power is required, such as would be needed for a small apartment building, a three-phase, four-wire wye system could be used. If the power requirement is very large, as for a machine shop, a three-phase, three-wire, delta system could be used. Figure 11-1 illustrates these systems.

The terms wye and delta refer to the windings used to produce the power. The basic patterns and symbols are shown in Figure 11-2.

Figure 11-3 shows how a single-phase, three-wire system can be used to produce different voltages. For a house, these connections are made in the service panel. Also shown are some possible combinations for a wye and delta system. Note that in each example, a neutral wire is always included.

TRANSFORMER WINDINGS

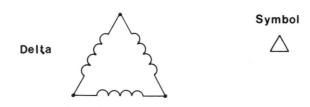

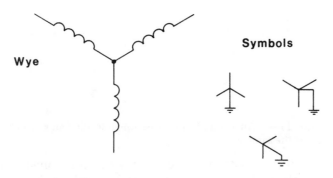

Figure 11-2 Transformer windings and their drawing symbols.

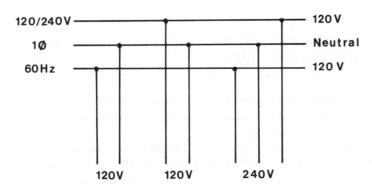

Figure 11-3 How a single-phase, three-wire system can be used to produce different voltages.

11-3 ONE-LINE DIAGRAMS

One-line diagrams are diagrams which graphically define the components and the relationship between the components for an electrical circuit. The components are arranged along a single vertical line which

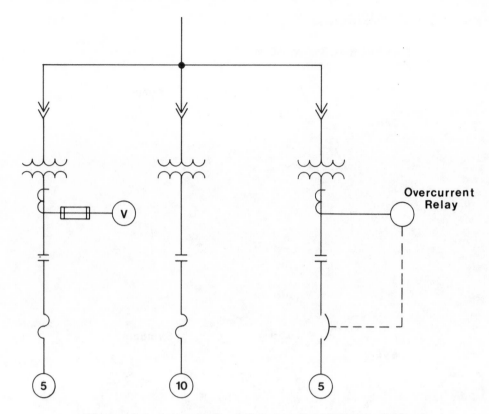

Figure 11-4 An example of three one-line diagrams which are part of the same circuit.

is read from top to bottom. Actually it is rare for a one-line diagram to consist of only a single line; most often, several one-line diagrams are drawn together to show the overall power distribution system rather than just a single circuit. Figure 11-4 shows a drawing which consists of several one-line diagrams.

11-4 HOW TO READ ONE-LINE DIAGRAMS

The basic symbols used to draw one-line diagrams are presented in Figure 11-5. Figure 11-6 shows a one-line diagram which has all symbols and notations labeled.

We see in Figure 11-6 that the incoming power is 480 volts, 3-phase, 60 Hz. The power passes through the connection and down the line which represents the circuit. The power then passes through a transformer which has a delta winding on one side and a grounded wye winding on the other. The resulting current is 120, 1ϕ, 60 Hz. Next an ammeter is placed on the line. The power rating and range of the ammeter are defined. An overload protection device is located in the ammeter line to protect the ammeter from damage.

The main circuit then continues through a circuit breaker, starter switch, another overload protection device, and finally into the 5-horsepower electrical motor at the end of the line.

It should be noted that the labeling on a one-line diagram is an important aspect of the diagram. The draftsman should always carefully check *all* labeling used and then have the design engineer double-check.

SYMBOLS for ONE LINE DIAGRAMS

NAME	SYMBOL

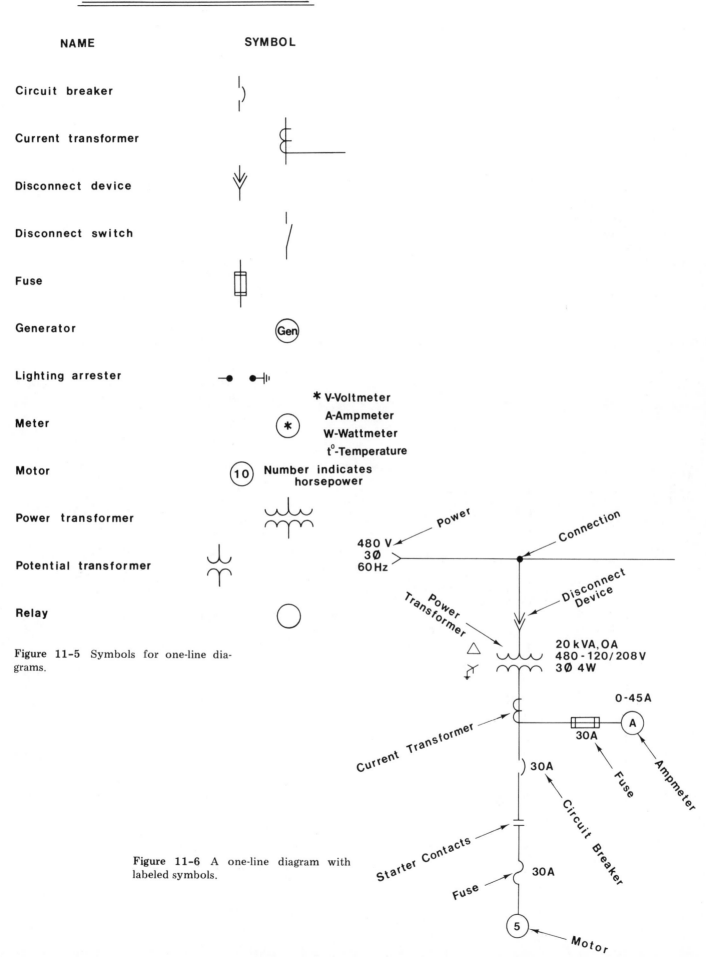

Circuit breaker

Current transformer

Disconnect device

Disconnect switch

Fuse

Generator — Gen

Lighting arrester

Meter — * V-Voltmeter / A-Ampmeter / W-Wattmeter / t°-Temperature

Motor — 10 — Number indicates horsepower

Power transformer

Potential transformer

Relay

Figure 11-5 Symbols for one-line diagrams.

Figure 11-6 A one-line diagram with labeled symbols.

Power

480 V
3Ø
60 Hz

Connection

Disconnect Device

Power Transformer

20 kVA, OA
480 - 120/208 V
3Ø 4W

Current Transformer

0-45A

A

30A

Fuse

Ampmeter

30A — Circuit Breaker

Starter Contacts

Fuse — 30A

5 — Motor

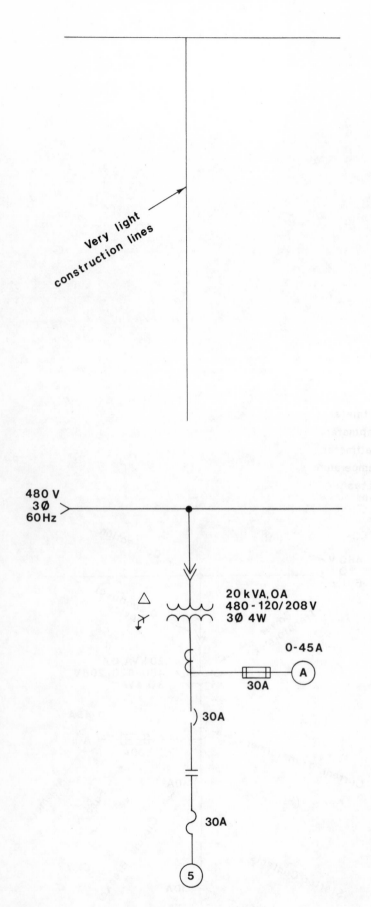

Very light construction lines

480 V
3Ø
60 Hz

20 kVA, OA
480 - 120/208 V
3Ø 4W

0-45A

30A

30A

30A

5

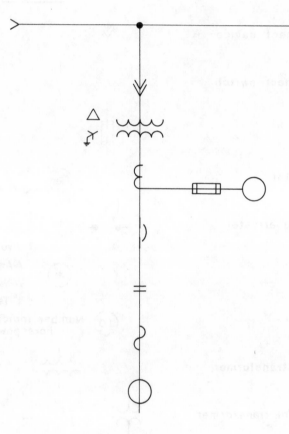

Figure 11-7 How to draw a one-line diagram.

11-5 HOW TO DRAW ONE-LINE DIAGRAMS

To draw a one-line diagram use the following procedure. The procedure is illustrated in Figure 11-7.

1. Prepare a freehand sketch of the entire circuit and check all symbols and labeling.

2. Draw very light horizontal lines to represent the power source and a vertical line as shown.

3. Add all symbols for the various components.

4. Add guide lines for labeling.

5. Darken all lines to their final color and configuration. Add labeling.

11-6 LADDER DIAGRAMS

Ladder diagrams are a type of diagram used to define industrial control circuits. They derive their name from the fact that they resemble ladders — two long parallel vertical lines connected by a series of horizontal runners. Figure 11-8 is an example of a ladder diagram.

Ladder diagrams are most commonly used to draw the circuitry required to activate and deactivate electrical motors and coils. These components are in turn used to operate elevators, heating and air conditioning systems, machines, subway systems, etc.

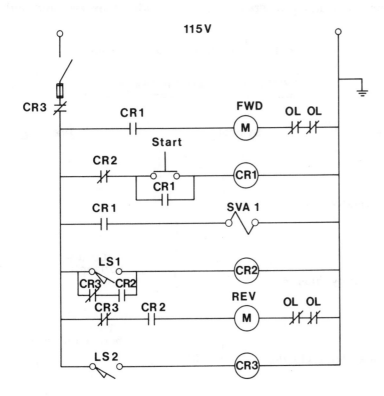

Figure 11-8 An example of a ladder diagram.

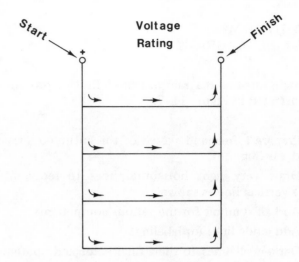

Figure 11-9 Ladder diagrams are read from left to right and top to bottom.

11-7   HOW TO READ LADDER DIAGRAMS

Ladder diagrams are read from top to bottom and left to right as illustrated in Figure 11-9. Power enters the circuit at the upper left corner and flows downward and across the diagram. The vertical line furthest to the right is the return line.

The symbols and abbreviations used on ladder diagrams are presented in Figure 11-10. Not only are the symbols important when drawing a ladder diagram but the labeling of the symbols is equally important. Ladder diagrams usually include many switches and coils

Figure 11-10   Symbols for ladder diagrams.

SYMBOLS for LADDER DIAGRAMS

Name	Symbol	Drawing Info
Contacts		
Normally open	─╫─ or	
Normally closed	─╫╱─ or	60° to horizontal $\frac{1}{8}$
Time delay, closing	─╫─ TDC	75° to horizontal $\frac{1}{2}$

Name	Symbol	Drawing Info

Time delay, opening TDO

Coils or Solenoids or $\frac{1}{2}$ DIA   30° to horizontal

Disconnect device 60° to horizontal

Circuit breaker $\frac{1}{2}$ DIA

Fuse $\frac{1}{2}$

Lamps (indicating) *R-Red   W-White   G-Green   B-Blue $\frac{1}{2}$ DIA

Overload devices or $\frac{1}{4}$ DIA

Switches   open   closed

    General

    Knife $\frac{1}{2}$

    Limit   by eye

Figure 11-10 Continued.

Name	Symbol		Drawing Info

Figure 11-10 Continued.

which, from a symbol viewpoint, appear exactly the same. The only way to tell which switch is related to which coil is by the labeling. For example, in Figure 11-11 we see that the solenoid labeled CR1 activates switch CR1 and the coil labeled CR2 activates CR2.

Switches are always drawn in their "natural" positions, that is, in the position they are in when not activated by current. A normally open switch is one which is open unless closed by current and is therefore drawn in the open position.

Figure 11-11 has been prepared to demonstrate how to read ladder diagrams. It is a ladder diagram showing the circuitry needed to start and stop an electric motor. The current enters the circuit at the positive terminal and exits at the negative terminal. Each part of the diagram has been labeled so you can follow the current as it flows through the circuit. Note that the diagram as drawn, shows all switches in their natural positions and that there are no completed paths across the diagram. Only when the start switch is pressed is there a completed path which permits current to flow. Once current is flowing, and the CR1 coil is activated the CR1 switches close and the start switch may be released.

All circuits should include overload protection. This may be in the form of fuses, temperature or current sensitive switches. These devices are also drawn in their natural position, which in most cases is the closed position. The standard abbreviation for overload protection is OL.

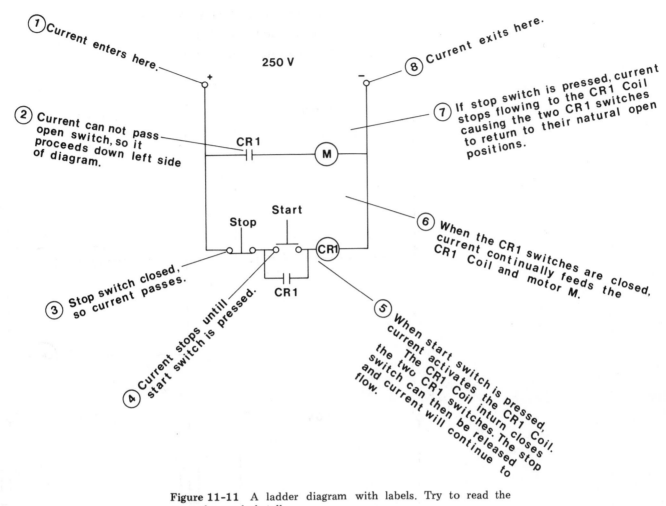

① Current enters here.

250 V

⑧ Current exits here.

② Current can not pass open switch, so it proceeds down left side of diagram.

⑦ If stop switch is pressed, current stops flowing to the CR1 Coil causing the two CR1 switches to return to their natural open positions.

CR1

M

Stop Start

③ Stop switch closed, so current passes.

⑥ When the CR1 switches are closed, current continually feeds the CR1 Coil and motor M.

CR1

④ Current stops untill start switch is pressed.

⑤ When start switch is pressed, current activates the CR1 Coil. The CR1 Coil inturn closes the two CR1 switches. The stop switch can then be released and current will continue to flow.

Figure 11-11 A ladder diagram with labels. Try to read the story the symbols tell.

11-8 HOW TO DRAW A LADDER DIAGRAM

The procedure used to create ladder diagrams is as follows (see Figure 11-12):

1. Make a freehand sketch of the entire circuit. A freehand sketch is helpful for two reasons: It will enable you to work out in advance the entire circuit, thereby reducing your chances of making an error in the final drawing, and it enables you to judge the approximate space requirements of the diagram. Remember, it is much easier to change a sketch than a final drawing.

2. Draw the two vertical lines which represent power in/power out and add any motors.

There is no standard distance between the vertical lines. Use your sketch to approximate the distance.

3. Add coils and control relay switches.
4. Add stop and start switches.
5. Add safety protection devices.

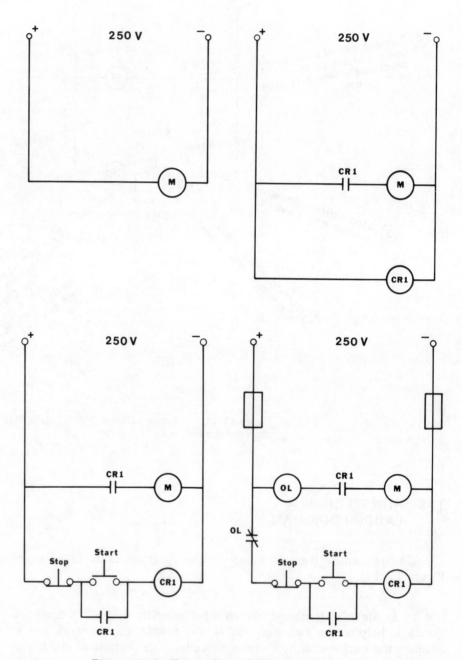

Figure 11-12   How to draw a ladder diagram.

Figure 11-13 is basically the same circuit as shown in Figure 11-12 except this time, the circuit has been redrawn to include a 115-volt control circuit. This means a transformer must be added as shown. Note how the 115-volt portion of the circuit is drawn using thinner lines than those drawn to represent the 220-volt portion. This

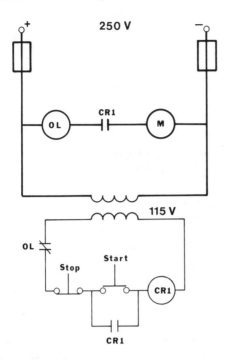

Figure 11-13 The same circuit as shown in Fig. 11-12 except this circuit uses a 115V control circuit.

is common practice; thicker lines represent higher voltages and thinner lines represent lower voltages.

11-9 RISER DIAGRAMS

Riser diagrams are drawings which show the wiring paths of a building's electrical system up to, but not including, the branch circuits. Figure 11-14 is an example of a riser diagram which shows the between floor wiring paths of a security system. Note that all lines drawn are uniform in intensity and thickness and that all lettering is as outlined in Figure 1-9. The size of the blocks may be varied according to individual needs, but the distance between floor lines is generally kept equal.

11-10 HOW TO READ A RISER DIAGRAM

Riser diagrams are read by starting at the power input point and then following each wiring path to its destination. In Figure 11-14 we see that the power enters the building in the basement. It then proceeds to the first Security Center Console and from there to the individual stations.

The legend in the lower right-hand corner of the drawing defines all the symbols used. All symbols should be clearly defined even if their meaning seems to be obvious.

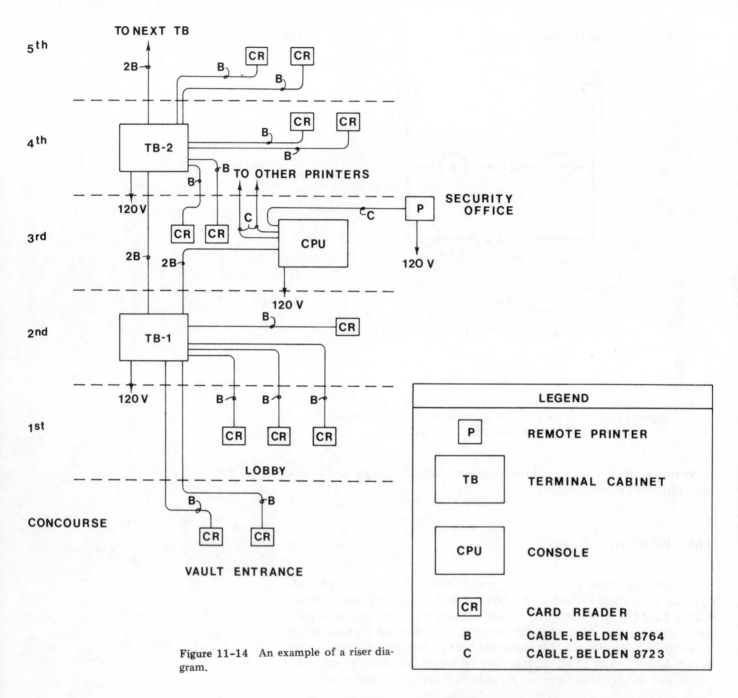

Figure 11-14 An example of a riser diagram.

LEGEND

P	REMOTE PRINTER
TB	TERMINAL CABINET
CPU	CONSOLE
CR	CARD READER
B	CABLE, BELDEN 8764
C	CABLE, BELDEN 8723

Each floor must be labeled and separated by a horizontal line. Special or large important components such as the Security Center Console and the Main Power Console are labeled on the drawing as shown. They should not be abbreviated.

11-11 HOW TO DRAW A RISER DIAGRAM

Figure 11-15 illustrates how to create a riser diagram. The diagram was originally presented in sketch form as shown in Figure 11-16. The procedure used is as follows.

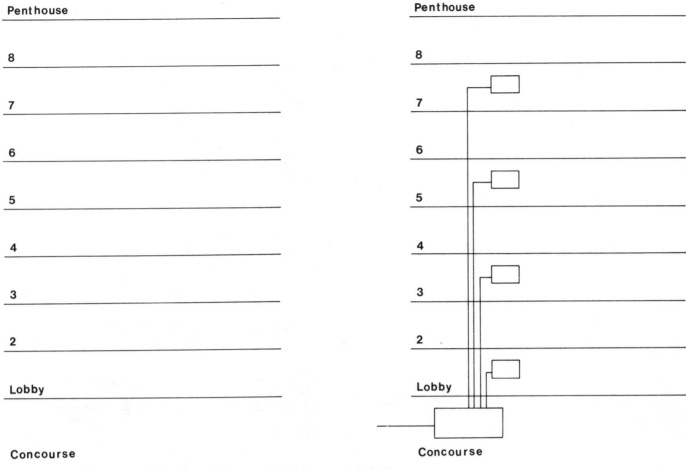

Figure 11-15   How to create a riser diagram.

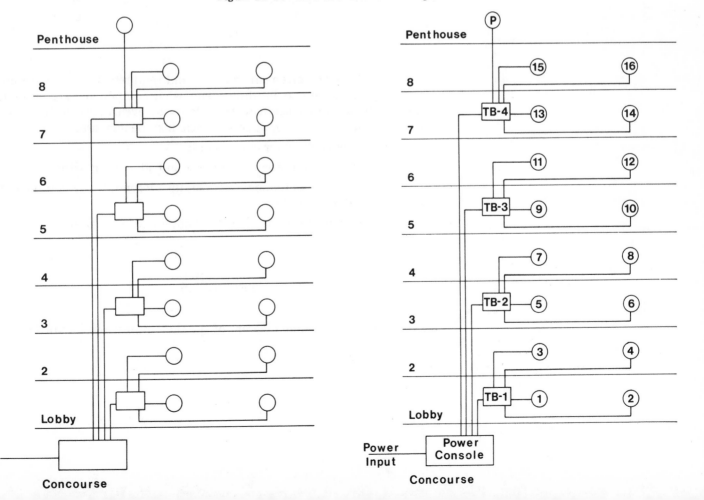

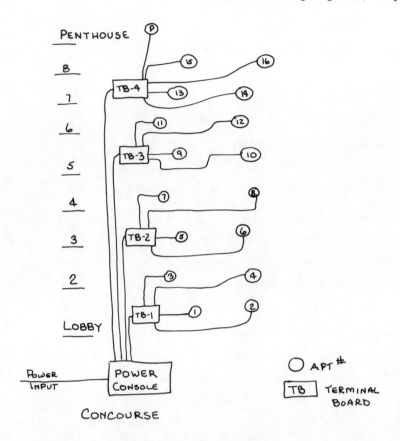

Figure 11-16 A freehand sketch of a riser diagram from which Fig. 11-15 was prepared.

1. Study the given information and requirements to determine how many floors are to be drawn and what components are needed. Draw the basic floor pattern so that floors are evenly spaced. If a floor has a large number of components, its size may be increased.

2. Draw in all major components.

3. Draw in all minor components and add wiring paths.

4. Label all components and set up a legend. Every component used must be defined in the legend.

5. Darken all lines to their final color and configuration.

PROBLEMS

11-1   Prepare a one line diagram of the sketch shown in Figure 11-17.

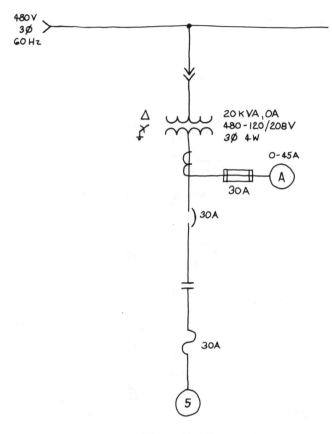

Figure 11-17

11-2 Prepare a one line diagram of the sketch shown in Figure 11-18.

Figure 11-18

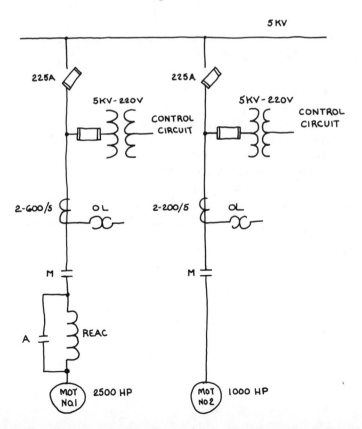

11-3   Prepare a one line diagram of the sketch shown in Figure 11-19.

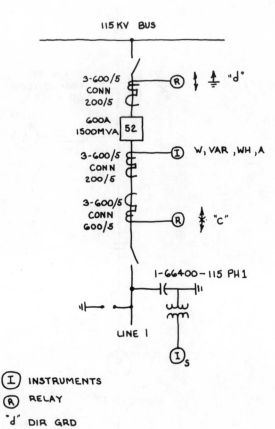

Figure 11-19

11-4   Redraw the ladder diagram shown in Figure 11-20 and substitute the following symbols:

1. Fuse L1
2. Fuse L2
3. Contact, normally open, CR1
4. A motor M
5. An overload device, OL
6. Transformer 115 V to 60 V
7. An overload device, OL
8. A start/stop sequence which includes a start button, stop button, a coil labeled CR1, and a normally open switch labeled CR1
9. An overload device, OL

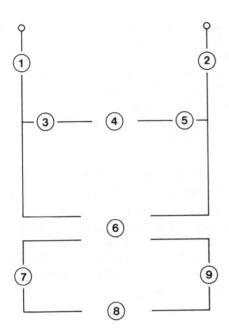

Figure 11-20

11-5 Figure 11-21 shows a drill press which has been rigged for automatic operation. The operator positions the workpiece under the drill and pushes a start button. The drill then follows the sequence listed

Figure 11-21

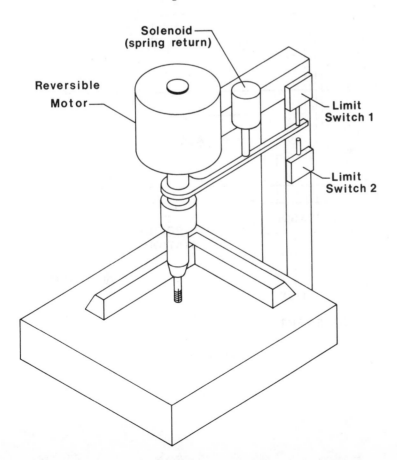

1. Motor starts — forward direction.
2. Solenoid activates and pushes drill downward.
3. Limit switch 2 is pressed.
4. Solenoid deactivates and returns (spring loaded return).
5. Motor stops and then reverses.
6. Limit switch 1 is pressed.
7. The motor stops and entire system is stopped.
8. All current is shut off.

Prepare a ladder diagram which represents the drilling sequence. Be sure to include overload and safety protection devices.

11-6   This problem is exactly like Problem 11-5 except in this case two more solenoids are added as clamping devices. These solenoids will press the workpiece against the existing guide rails shown and hold the workpiece during drilling. After the drilling is complete, the solenoids will release the workpiece.

11-7   Redraw the riser diagram shown in Figure 11-22 and add fourth floor with two more alarms.

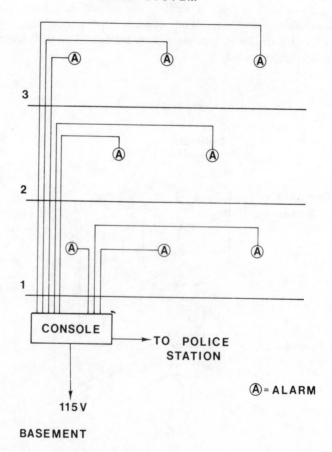

Figure 11-22

11-8 Redraw the riser diagram shown in Figure 11-23.

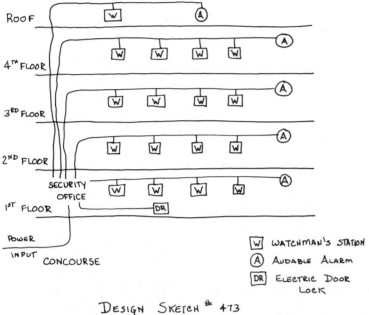

RISER DIAGRAM — WATCHMAN'S TOUR

ROOF

4TH FLOOR

3RD FLOOR

2ND FLOOR

SECURITY OFFICE

1ST FLOOR

POWER INPUT

CONCOURSE

W WATCHMAN'S STATION
A AUDABLE ALARM
DR ELECTRIC DOOR LOCK

DESIGN SKETCH # 473
JOHNSTON WAREHOUSE BLD

Figure 11-23

11-9　Redraw the riser diagram shown in Figure 11-24.

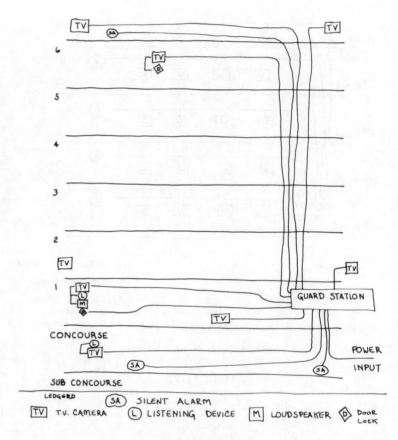

Figure 11-24

APPENDIXES

A DIGITAL READOUT LETTERS

DIGITAL READOUTS

Display Shape

Basic Letter Shapes

0 1 2 3 4

5 6 7 8 9

Variations

8 3 4

8 3 4

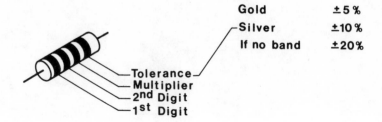

B RESISTOR COLOR CODES

Gold	±5%
Silver	±10%
If no band	±20%

Tolerance
Multiplier
2nd Digit
1st Digit

Color	1st Digit	2nd Digit	Multiplier
Black	0	0	1
Brown	1	1	10
Red	2	2	100
Orange	3	3	1,000
Yellow	4	4	10,000
Green	5	5	100,000
Blue	6	6	1,000,000
Violet	7	7	10,000,000
Gray	8	8	100,000,000
White	9	9	1,000,000,000
Gold	—	—	.1
Silver	—	—	.01

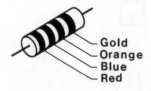

Gold
Orange
Blue
Red

2600 ohms ± 5%

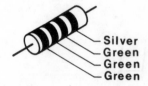

Silver
Green
Green
Green

550,000 ohms ±10%

WIRE AND
C SHEET METAL GAGES

WIRE AND SHEET METAL GAGES			
Gage	Thickness	Gage	Thickness
000 000	0.5800	18	0.0403
00 000	.5165	19	.0359
0 000	.4600	20	.0320
000	.4096	21	.0285
00	.3648	22	.0253
0	.3249	23	.0226
1	.2893	24	.0201
2	.2576	25	.0179
3	.2294	26	.0159
4	.2043	27	.0142
5	.1819	28	.0126
6	.1620	29	.0113
7	.1443	30	.0100
8	.1285	31	.0089
9	.1144	32	.0080
10	.1019	33	.0071
11	.0907	34	.0063
12	.0808	35	.0056
13	.0720	36	.0050
14	.0641	37	.0045
15	.0571	38	.0040
16	.0508	39	.0035
17	.0453	40	.0031

D PAPER SIZES

PAPER SIZES

Size	Dimensions
A	$8\frac{1}{2}$ x 11
B	11 x 17
C	17 x 22
D	22 x 34
E	34 x 44
J	Roll Size

E WIRE COLOR CODES

COLOR CODES		
COLOR	LETTER CODE	NUMBER CODE
Black	BK	0
Brown	BR	1
Red	R	2
Orange	O	3
Yellow	Y	4
Green	GR	5
Blue	BL	6
Violet	V	7
Gray	GY	8
White	W	9

INDEX